Sets and groups

A first course in algebra

J. A. Green

Professor of Mathematics
University of Warwick

Routledge & Kegan Paul
London and New York

First published in 1965
Second, revised edition 1988 by
Routledge & Kegan Paul Ltd
11 New Fetter Lane, London EC4P 4EE

Published in the USA by
Routledge & Kegan Paul Inc.
in association with Methuen Inc.
29 West 35th Street, New York, NY 10001

Set in 10/12pt Times
by Cotswold Typesetting Ltd, Gloucester
and printed in Great Britain
by Richard Clay Ltd,
Bungay, Suffolk

Library of Congress Cataloging in Publication Data

Green, J. A. (James Alexander)
    Sets and groups.

    Includes index.
    1. Groups, Theory of. 2. Set theory.
I. Title.
QA171.G68   1987   512   86–33887

British Library CIP Data also available
ISBN 0-7102-0557-0(c)
        0-7102-1227-5(p)

Contents

Preface to first edition

Mathematics is attractive and useful and limited because it attempts a precise expression of its ideas; people learn it in the hope of benefiting both from the ideas themselves and from the way they are expressed. But mathematics is always escaping from its own limits, and it produces new ideas and new language which represent a continual challenge to mathematical teaching.

The first three chapters of this book describe sets, relations and mappings; these are the basic ideas in terms of which most modern mathematics can be expressed. They are certainly 'abstract' ideas, but they are not difficult to grasp, because they come from notions which most people are used to from everyday experience. What is characteristic of modern mathematics is that it uses these few ideas over and over again, to build up theories which cover a great variety of situations. The rest of the book is an introduction to one such theory, namely the theory of groups.

The book is set out so that each new idea is described in its general form, and then illustrated by examples. These examples form a large part of the exposition, and require only a small amount of traditional mathematics to understand. There are also numerous exercises for the reader, at the end of each chapter.

It is a pleasure to thank my friends Drs W. Ledermann, L. W. Morland and S. Swierczkowski, who each read and commented valuably on the manuscript.

J. A. GREEN

Preface to revised edition

In preparing this revised edition of *Sets and Groups* I have made some changes to the original text, and have added a substantial amount of new material. The main changes to the original text are that 'mappings' are now called 'maps', and that they are written 'on the left' instead of 'on the right'. It is hoped in this way to keep up with current mathematical fashion on these important matters.

The introduction of new material – particularly the two new, long, chapters 8 and 9 – calls for some explanation. The original *Sets and Groups*, like the other books in the 'Library of Mathematics' series, was short and inexpensive; it was envisaged as a preparatory text for further study at first-year university or college level, particularly in algebra. Unfortunately book production costs have risen so fast in recent years that it is no longer possible to offer *Sets and Groups*, in its original form, at a price attractive to students. It seemed to me that the best solution was to enlarge the book so that it would cover most of the algebra likely to appear in first-year courses in British universities; to make it in fact a first course in modern algebra, suitable for specialist as well as non-specialist mathematics students.

To achieve this, I have added material on groups to chapter 7, and I have written new chapters on rings and fields (chapter 8) and on vector spaces and matrices (chapter 9). I have retained, without essential change, the early chapters of *Sets and Groups*, which introduce the reader to modern mathematical language, and to algebraic structures (sets with binary operations).

This book is written for all who are beginning to study algebra at university or college level, and it is expected that the reader will select the parts which he or she needs. To facilitate this, the last two chapters are largely independent of each other, and of the latter half (sections 7.5–7.8) of chapter 7.

January 1986   *J. A. Green*

1 Sets

1.1 Sets

Elementary mathematics is concerned with counting and measurement, and these are described in terms of *numbers*. Modern 'abstract' mathematical disciplines start from the more fundamental idea of a *set*.

A set is simply a collection of things, which are called the *elements* or *members* of the set. We think of a set as a single object in its own right, and often denote sets by capital letters A, B, etc.

If A is a set, and if x is an element (or member) of A, we say x belongs to A. The symbol $\in$ is used for 'belongs to' in this sense, so that

$$x \in A$$

is a notation for the statement 'x is an element of A' or 'x belongs to A'. The negation of this, namely the statement 'x is not an element of A', is denoted $x \notin A$.

A set is determined by its elements, which means that a set A is fully described by describing all the elements of A. Sets A, B are defined to be *equal* (and we write $A = B$) if and only if they have the same elements, i.e. every element of A is an element of B, and every element of B is an element of A.

The notion of a set is very general, because there is virtually no restriction on the nature of the things which may be elements of a set. We give now some examples of sets.

Example 1. Let Z denote the set of all integers, i.e. the elements of Z are the integers (including zero and negative integers)

$$\ldots, -2, -1, 0, 1, 2, \ldots.$$

As examples of the $\in$ notation, we could write $0 \in Z$, $-4 \in Z$ and $\frac{1}{2} \notin Z$.

Example 2. Now let R denote the set of *all* real† numbers. Then for example 0, -4, $\frac{1}{2}$, $\sqrt{2}$, are some of the elements of R. A good way to visualize this set is to represent its elements by points on a straight line, as in co-ordinate geometry. A point on the line is chosen, arbitrarily, to represent the number 0, putting positive x to the right, and negative x to the left of 0 in the usual way.

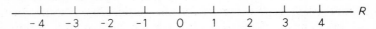

Throughout this book, the letters Z and R will be reserved for the two sets just described. See also the list of symbols on p. 247.

Example 3. Let P be the set of all people who are living at this moment. Then the reader of this page presumably belongs to P, but Julius Caesar does not, nor does the integer -4.

Example 4. Sometimes it is possible and convenient to display the elements of a set between brackets { }, for example $S = \{a, b, c\}$ means, S is the set whose elements are a, b and c. The order in which the elements are written does not matter, for example $\{b, a, c\}$ is the same set as S, since it has the same elements.

Example 5. A set A may have only one element, a, say, so that $A = \{a\}$. This 'one-element set' $\{a\}$ is logically different from a itself, because, for example, it is true that $a \in \{a\}$, but not that $a \in a$.

Finite and infinite sets

A set A which has only a finite number of elements is called a *finite set*, otherwise A is an *infinite set*. The number of elements in a finite set A is called the *order* of A and is denoted $|A|$.

Example 6. The sets Z and R are infinite. The set S of Example 4 is finite and $|S| = 3$. The set P of Example 3 is finite, but it is unlikely that its order will ever be known exactly!

---

†Ordinary numbers are often called *real* numbers, to distinguish them from complex numbers.

A notation for sets

The following type of notation is often used to describe a set:

$Z = \{x \mid x \text{ is an integer}\}.$

This is read 'Z is the set of all x such that x is an integer', which means the same as 'Z is the set of all integers'. In this notation, the symbol (such as x) on the left of the vertical line $\mid$ stands for a typical element of the set, while on the right of the line is a statement about this typical element (such as the statement 'x is an integer') which serves to determine the set exactly.

Example 7. $Z(n) = \{x \in Z \mid 1 \leq x \leq n\}$ means, $Z(n)$ is the set of all x belonging to Z (i.e. of all *integers* x) such that $1 \leq x \leq n$. In other words, $Z(n)$ is the set whose elements are $1, 2, \ldots, n$. For example $Z(1) = \{1\}$, $Z(2) = \{1, 2\}$, $Z(3) = \{1, 2, 3\}$.

Example 8. $R^+ = \{x \in R \mid x > 0\}$ means, R^+ is the set of all *positive* real numbers.

1.2 Subsets

DEFINITION. Let A be any set. Then a set B is called a *subset* of A if every element of B is an element of A.

The notations $B \subseteq A$ and $A \supseteq B$ are both used to mean 'B is a subset of A'.† According to our definition, a set A is always a subset of itself, $A \subseteq A$. Any subset B of A which is not equal to A, is called a *proper* subset of A.

If $A \subseteq B$ and also $B \subseteq A$, then $A = B$, because then every element of A is an element of B and every element of B is an element of A.

Example 9. Z is a subset of R. The set B of all British people is a subset of the set P of all people. For any positive integer n, $Z(n)$ is a subset of Z. All these are proper subsets.

Example 10. Figure 1 is supposed to represent, diagrammatically, the relation $B \subseteq A$. A is represented by the set of all points inside the large

---

†The symbol $\subset$ is often used instead of $\subseteq$.

circle, and B by the set of all points inside the small one. (Figures of this sort are sometimes called 'Venn diagrams'.)

Example 11. The notation $B \subseteq A$ is intended to recall the notation $b \le a$ (meaning 'b is less than or equal to a') for numbers. But there is here an important difference between sets and numbers, namely that if A, B are two sets, it can easily happen that neither $A \subseteq B$ nor $B \subseteq A$ (for example, take $A = Z$ and $B = P$). But for any numbers a, b it is always true that $a \le b$ or $b \le a$.

The implication sign $\Rightarrow$

The sign $\Rightarrow$ is often used between two statements as an abbreviation for 'implies'. For example, if B is a subset of A we can write

$x \in B \Rightarrow x \in A,$

meaning 'x belongs to B implies x belongs to A', which is just another way of stating our definition 'Every element of B is an element of A'.

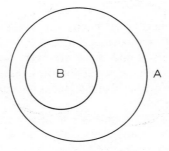

Figure 1

If two statements are such that each implies the other, we write $\Leftrightarrow$ between them. This sign can be read as 'implies and is implied by', or less clumsily 'if and only if'. Thus the condition that sets A, B should be equal may be written

$x \in A \Leftrightarrow x \in B.$

1.3 Intersection

DEFINITION. If A and B are sets, then the *intersection* $A \cap B$ of A and B is the set of all elements which belong to both A and B.

The notations introduced in the last two sections give us two other, but entirely equivalent, ways of putting this definition. First we might write

$$A \cap B = \{x \mid x \in A \text{ and } x \in B\},$$

i.e. '$A \cap B$ is the set of all x such that $x \in A$ and $x \in B$'. Alternatively the condition that a given thing x belongs to $A \cap B$ can be expressed

$$x \in A \cap B \Leftrightarrow x \in A \text{ and } x \in B.$$

Example 12. If $S = \{a, b, c\}$ and $T = \{c, e, f, b\}$, then $S \cap T = \{b, c\}$. If B is the set of all British people, and C is the set of all blue-eyed people, then $B \cap C$ is the set of all blue-eyed, British people.

Example 13. Figure 2 shows $A \cap B$ as the shaded part which is in common to the circles representing A and B.

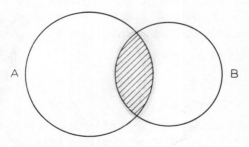

Figure 2

For any sets A, B it is clear that $A \cap B = B \cap A$. Also $A \cap B$ is a subset of both A and B. In fact if Y is any set such that $Y \subseteq A$ and also $Y \subseteq B$, then $Y \subseteq A \cap B$ (because every element y of Y is an element of both A and B, hence $y \in A \cap B$). Thus $A \cap B$ is the 'largest' set which is a subset of both A and B.

Disjoint sets. The empty set

If sets A, B have no element in common, they are called *disjoint*. For example the set Z and P of section 1.1 are disjoint, because there is no integer which is also a person.

If A, B are disjoint, then $A \cap B$ is not really defined, because it has no elements. For this reason we introduced a conventional *empty set*, denoted $\varnothing$, to be thought of as a 'set with no elements'. Of course this is a set only by courtesy, but it is convenient to allow $\varnothing$ the status of a set. If A, B are disjoint, we write $A \cap B = \varnothing$.

The empty set is counted as a finite set, with order $|\varnothing| = 0$. We regard $\varnothing$ as a subset of any set A, an assumption which does not contradict our previous notion of subset. For the statement '$\varnothing \subseteq A$' should mean 'if $x \in \varnothing$, then $x \in A$'. Since there is no x such that $x \in \varnothing$, this last statement is never contradicted.

1.4 Union

DEFINITION. If A and B are sets, then the *union* $A \cup B$ of A and B is the set of all elements of A, together with all elements of B. Alternative expressions of this definition are

$$A \cup B = \{x \mid x \in A \text{ or } x \in B\},$$

or

$$x \in A \cup B \Leftrightarrow x \in A \text{ or } x \in B,$$

where in both statements, the word 'or' is used in a sense which includes 'and', i.e. $A \cup B$ is the set of all elements which belong to A or to B, *including* those which belong to both A and B. Thus $A \cap B$ is always a subset of $A \cup B$ (see Figure 2, where $A \cup B$ is the whole region bounded by the two circles, including the shaded part).

Example 14. Let S, T, B, C be as in Example 12. Then $S \cup T = \{a, b, c, e, f\}$, and $B \cup C$ is the set of all people who are either British, or blue-eyed, or both.

Example 15. For any sets A, B it is clear that $A \cup B = B \cup A$. Also both A and B are subsets of $A \cup B$. In fact if X is any set such that $A \subseteq X$ and $B \subseteq X$, then $A \cup B \subseteq X$ (because every element x of $A \cup B$ is an element of A or else an element of B, and in either case $x \in X$). Thus $A \cup B$ is the 'smallest' set in which both A and B are subsets.

Example 16. Let A, B be finite sets. Then

$$|A \cup B| = |A| + |B| - |A \cap B|.$$

For we can count the elements of $A \cup B$, by first counting the $|A|$ elements of A, and then the $|B|$ elements of B; but then we have counted twice each of the elements of $A \cap B$, so we must subtract $|A \cap B|$ from $|A| + |B|$ to get the total.

1.5 The algebra of sets

The operations $A \cap B$, $A \cup B$ are in some ways like the operations ab, $a + b$ with numbers. In the 'algebra of sets'† we can prove identities, as in ordinary algebra. Two simple examples of such *set-theoretic identities* are $A \cap B = B \cap A$, and $A \cup B = B \cup A$, both of which hold for all sets A, B. These resemble the algebraic identities $ab = ba$ and $a + b = b + a$, which hold for all numbers a, b. The identity proved in Example 17, below, resembles the algebraic identity $a(b + c) = ab + ac$.

The analogy between sets and numbers does not go very far. For example $A \cap A = A$ is an identity for sets, but its algebraic counterpart $aa = a$ is not an identity, since it does not hold for all numbers a.

There is an interesting *duality principle* for sets, according to which any identity involving the operation $\cap$, $\cup$ (and no others) remains valid if the symbols $\cap$, $\cup$ are interchanged throughout (see Example 18 for an illustration of this). But there is no corresponding 'duality' between ordinary multiplication and addition. For example, if we interchange addition and multiplication in the identity $a(b + c) = (ab) + (ac)$ (we have put in some extra parentheses to make this easier to do) we get $a + (bc) = (a + b)(a + c)$, and this is not a correct algebraic identity.

---

†Invented by G. Boole (1815–64).

Example 17. To prove that $A\cap(B\cup C)=(A\cap B)\cup(A\cap C)$, for all sets A, B, C.

Proof. The reader should check each step below, remembering that $\Leftrightarrow$ means 'if and only if': $x\in A\cap(B\cup C)\Leftrightarrow x\in A$ and $x\in B\cup C\Leftrightarrow x\in A$ and ($x\in B$ or $x\in C)\Leftrightarrow(x\in A$ and $x\in B)$ or ($x\in A$ and $x\in C)\Leftrightarrow x\in A\cap B$ or $x\in A\cap C\Leftrightarrow$ $x\in(A\cap B)\cup(A\cap C)$. This proves that every element x of $A\cap(B\cup C)$ is an element of $(A\cap B)\cup(A\cap C)$, and conversely. Hence the two sets are equal.

Example 18. The 'dual' of the last identity is $A\cup(B\cap C)=(A\cup B)\cap(A\cup C)$. To prove this, we can 'dualize' the proof just given, by interchanging the symbols $\cap$, $\cup$ and also the words 'and', 'or', throughout. The reader will find that this gives automatically a proof of this new identity.

1.6 Difference and complement

DEFINITION. If A and B are sets, then the *difference set* $A-B$ is the set of those elements of A which do not belong to B.†

Alternative expressions of this definition are

$$A-B=\{x|x\in A \text{ and } x\notin B\},$$

or

$$x\in A-B\Leftrightarrow x\in A \text{ and } x\notin B.$$

Complement. If B is a subset of A, then $A-B$ is sometimes called the *complement* of B in A. Clearly B is the complement of $A-B$ in A (see Figure 4).

Example 19. Figure 3 shows the relative situations of the three sets $A-B$, $A\cap B$ and $B-A$. If $A\subseteq B$ it is clear that $A-B=\varnothing$, in particular $A-A=\varnothing$.

Example 20. For any sets A and B, $A-B=A-(A\cap B)$. If A is finite and B is a subset of A, then $|A-B|=|A|-|B|$. This formula shows why the sign $-$ is used for difference of sets, but it should be remembered that $A-B$ is defined, even when B is not a subset of A.

Example 21. Sets A_1, A_2, A_3 are represented in Figure 5 by the interiors of

---

†The notation $A\backslash B$ is often used, instead of $A-B$, to denote the difference set. But in this book $A\backslash B$ will have a different meaning (see Chapter 6, section 6.1).

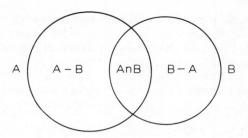

Figure 3

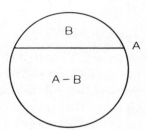

Figure 4

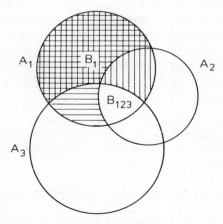

Figure 5

the circles shown. $A_1 - A_2$ is the region shaded horizontally, and $A_1 - A_3$ is shaded vertically. Their intersection is B_1, and the diagram shows that this is also $A_1 - (A_2 \cup A_3)$. This suggests (but does not prove!) the identity.

$$(A_1 - A_2) \cap (A_1 - A_3) = A_1 - (A_2 \cup A_3).$$

We prove this formally, for any sets A_1, A_2, A_3, as follows:

$x \in (A_1 - A_2) \cap (A_1 - A_3) \Leftrightarrow x \in A_1 - A_2$ and $x \in A_1 - A_3 \Leftrightarrow (x \in A_1$ and $x \notin A_2)$ and $(x \in A_1$ and $x \notin A_3) \Leftrightarrow x \in A_1$ and x belongs to neither A_2 nor $A_3 \Leftrightarrow x \in A_1$ and $x \notin A_2 \cup A_3 \Leftrightarrow x \in A_1 - (A_2 \cup A_3)$.

1.7 Pairs. Product of sets

In ordinary plane co-ordinate geometry, a point is described by two co-ordinates x, y, that is, the point is described by a *pair* (x, y) of numbers.

The pair (x, y) is not the same as the *set* $\{x, y\}$ which has x and y as its elements. In a pair, the order of the terms is essential (in fact the term *ordered pair* is often used to emphasize this). For example, $(1, 2)$ and $(2, 1)$ represent different points in the plane (see Figure 6), but the sets $\{1, 2\}$ and $\{2, 1\}$ are the same, because they have the same elements. Again, we may have a pair (x, y) with $x = y$, for example $(0, 0)$ or $(1, 1)$.

We have mentioned co-ordinate geometry as an illustration, but now we shall formulate the idea of a pair in general terms. If x and y are any things (they need not be numbers, and they need not be different) let us use the symbol (x, y) to denote the pair consisting of x and y in that order. Pairs (x, y) and (x', y') are considered *equal* if and only if $x = x'$ *and* $y = y'$. x is called the first, and y the second, *component* of (x, y).

DEFINITION. If A and B are sets, then the *product* (sometimes called *Cartesian product*) $A \times B$ of A and B is the set of all pairs (x, y) such that $x \in A$ and $y \in B$.

Example 22. Let $A = \{1, 2, 3\}$ and $B = \{a, b\}$. Then $A \times B$ has six elements, *viz.* $(1, a), (2, a), (3, a), (1, b), (2, b), (3, b)$. $B \times A$ also has six elements, obtained by reversing these, *viz.* $(a, 1), (a, 2), (a, 3), (b, 1), (b, 2), (b, 3)$. If A, B are any finite sets, then $|A \times B| = |A||B|$, because there are $|A|$ ways of choosing the first component of an element (x, y) of $A \times B$, and, for each such choice, $|B|$ ways of choosing the second component. This is the reason for using the sign $\times$ for product of sets.

Example 23. If A is any set, then $A \times A$ is the set of all pairs (x, y), where both x and y belong to A. For example, if R is the set of all real numbers, then $R \times R$ is exactly the set of all the pairs (x, y) which represent points in plane co-ordinate geometry. Plane curves and other figures can be regarded as subsets of $R \times R$. For example the set $E = \{(x, y)|y = x^2\}$ is the subset of $R \times R$ represented on Figure 6 by the curve with equation $y = x^2$, because this is just the set of those points (x, y) such that this equation holds. The set shaded represents the subset $F = \{(x, y)|y \le x^2\}$ of $R \times R$.

Triples, n-tuples

The idea of a pair can be generalized. For example if x, y, z are any things (not necessarily distinct) we use the symbol (x, y, z) to denote the *triple* consisting of x, y, z in that order. Triples (x, y, z) and (x', y', z') are considered equal if and only if $x = x'$, $y = y'$ and $z = z'$. If A, B, C are sets, we denote by $A \times B \times C$ the set of all triples (x, y, z) such that $x \in A$, $y \in B$, $z \in C$.

Still more generally, we may take any positive integer n, and speak of the *n-tuple* $(x_1, x_2, \ldots, x_n)$ whose *components* $x_1, x_2, \ldots, x_n$ are some given things (not necessarily distinct); *n*-tuples $(x_1, x_2, \ldots, x_n)$ and $(y_1, y_2, \ldots, y_n)$ are considered equal if and only if $x_1 = y_2, x_2 = y_2, \ldots, x_n = y_n$. If $A_1, A_2, \ldots, A_n$ are any sets, we denote by $A_1 \times A_2 \times \ldots \times A_n$ the set of all *n*-tuples $(x_1, x_2, \ldots, x_n)$ such that $x_1 \in A_1, x_2 \in A_2, \ldots, x_n \in A_n$. In the case where all the sets $A_1, A_2, \ldots, A_n$ are equal to some given set A, we denote $A_1 \times A_2 \times \ldots \times A_n = A \times A \times \ldots \times A$ by A^n. An important case is the set R^n of all *n*-tuples $x = (x_1, x_2, \ldots, x_n)$ whose components $x_1, x_2, \ldots, x_n$ are all real numbers. These *n*-tuples are often called *n-vectors* (see section 9.2).

1.8   Sets of sets

Later in this book, we shall find that we often have to deal with a *set of sets*, that is, with a set whose elements are themselves sets.

Sometimes we shall use script capitals, such as $\mathscr{P}$, $\mathscr{S}$, etc. to denote such sets of sets. The only new feature is that we must be careful how we use the sign $\in$. If $\mathscr{S}$ is a set of sets, and if $A \in \mathscr{S}$, then A is itself a set. But an element, say x, of A is *not* usually an element of $\mathscr{S}$. Thus $x \in A$ and $A \in \mathscr{S}$, but in general $x \notin \mathscr{S}$.

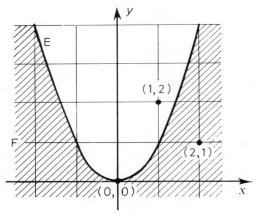

Figure 6

Example 24. Let $A = \{1, 2, 3, 4\}$, $B = \{a, 3, 4\}$, $C = \{2, a\}$. Then $\mathscr{S} = \{A, B, C\}$ is a set of sets. It has three elements A, B, C. The elements 1, 2, 3, 4, a of these sets are not elements of $\mathscr{S}$.

Example 25. If X is a given set, we shall always denote by $\mathscr{B}(X)$ the set of *all subsets of X*. In particular $\varnothing$ and X itself are elements of $\mathscr{B}(X)$. As an example take $X = \{x, y, z\}$. Then $\mathscr{B}(X)$ has eight elements $\varnothing$, $\{x\}$, $\{y\}$, $\{z\}$, $\{x, y\}$, $\{x, z\}$, $\{y, z\}$, $\{x, y, z\}$. Notice that $\{x\} \in \mathscr{B}(X)$ but $x \notin \mathscr{B}(X)$.

Intersection and union

Let $\mathscr{S}$ be any set of sets. We define the intersection $\cap \mathscr{S}$ to be the set of all objects x which belong to all of the sets A in $\mathscr{S}$, i.e.

$$\cap \mathscr{S} = \{x \mid x \in A, \quad \text{for all } A \in \mathscr{S}\}.$$

Similarly the union $\cup \mathscr{S}$ is defined to be the set of all objects x which belong to at least one set A in $\mathscr{S}$, i.e.

$$\cup \mathscr{S} = \{x \mid x \in A, \qquad \text{for at least one } A \in \mathscr{S}\}.$$

If $\mathscr{S}$ has only two members, $\mathscr{S} = \{A, B\}$, then these definitions reduce to our earlier definitions for $A \cap B$ and $A \cup B$ respectively, i.e. $\cap \{A, B\} = A \cap B$ and $\cup \{A, B\} = A \cup B$. If $\mathscr{S}$ is a finite set of sets, say $\mathscr{S} = \{A_1, \ldots, A_n\}$, we often write

$$\cap \mathscr{S} = A_1 \cap \ldots \cap A_n,$$

and

$$\cup \mathscr{S} = A_1 \cup \ldots \cup A_n.$$

We shall not, in fact, use the notations $\cap$ and $\cup$ very much in this book.

Example 26. Take $\mathscr{S} = \{A, B, C\}$ as in Example 24. Then $\cap \mathscr{S} = A \cap B \cap C = \varnothing$ and $\cup \mathscr{S} = A \cup B \cup C = \{1, 2, 3, 4, a\}$. Notice that $A \cap B \cap C$ is empty, although none of $A \cap B$, $B \cap C$, $A \cap C$ is.

Example 27. In Figure 5, $A_1 \cap A_2 \cap A_3$ is represented by the region B_{123}, and $A_1 \cup A_2 \cup A_3$ by the whole figure. Notice

$$(A_1 \cap A_2) \cap A_3 = A_1 \cap A_2 \cap A_3 = A_1 \cap (A_2 \cap A_3)$$

and the dual identities for unions.

Example 28. As example of an infinite set of sets take $\mathscr{S} = \{Z(n) \mid n = 1, 2, 3, \ldots\}$, i.e. the *elements of* $\mathscr{S}$ are the set $Z(1) = \{1\}$, $Z(2) = \{1, 2\}$, $Z(3) = \{1, 2, 3\}$, etc. (Notice that $\mathscr{S}$ is a *subset* of $\mathscr{B}(Z)$.) It is easy to see that $\cap \mathscr{S} = \{1\}$ and $\cup \mathscr{S}$ is the set of all positive integers.

Russell's paradox†

Serious difficulties occur if we allow the notion of set to be too general. For example it is undesirable to talk of 'the set U of all sets'. If such a set exists, it must, being a set, be a member of itself, $U \in U$. This is not in itself disastrous, but now consider Russell's famous set V of *all sets which are not members of themselves*. If V is

---

†Invented by Bertrand Russell in 1901.

not a member of itself, then by that fact it qualifies as member of V, i.e. it *is* then a member of itself. And the situation is no better if we yield to this reasoning and allow that V *is* a member of itself. For then V itself does not qualify as member of V, which only admits as members those sets which are not members of themselves, so we find that we have again proved the opposite of what we assumed. This logical impasse can be avoided by restricting the notion of set, so that 'very large' collections such as U or V or the 'collection of all things' are not counted as sets. However, this is done at some cost in simplicity, and in this book we shall do no more than keep to sets which appear to be harmless, and hope that paradoxes will not appear.

Exercises for Chapter 1

1 Describe in words the sets $A = \{x \in Z | 2 \leq x\}$, $B = \{x \in Z | x \leq 5\}$. Show that $A \cap B$ is finite and that $A \cup B = Z$.

2 If A and B are any sets, prove that $A \cap B = B$ if and only if $A \supseteq B$.

3 Prove the identities $(A \cap B) \cup A = A$ and $(A \cup B) \cap A = A$.

4 Let A_1, A_2, A_3 be finite sets, and write $a_i = |A_i|$, $a_{ij} = |A_i \cap A_j|$ and $a_{123} = |A_1 \cap A_2 \cap A_3|$. Prove that $|A_1 \cup A_2 \cup A_3| = a_1 + a_2 + a_3 - a_{12} - a_{13} - a_{23} + a_{123}$.

5 With the notation of Exercise 4, suppose that $a_1 = 10$, $a_2 = 15$, $a_3 = 20$, $a_{12} = 8$ and $a_{23} = 9$. Prove that the only values which $|A_1 \cup A_2 \cup A_3|$ could have are 26, 27 or 28.

6 Prove the identity $(A_1 - A_2) \cup (A_1 - A_3) = A_1 - (A_2 \cap A_3)$.

7 For any sets A, B define the 'symmetric sum' $A \oplus B$ to be the set $(A \cup B) - (A \cap B)$. Draw a diagram to represent $A \oplus B$. Prove the identities (i) $A \oplus B = B \oplus A$; (ii) $A \oplus A = \varnothing$; (iii) $(A \oplus B) \oplus C = A \oplus (B \oplus C)$; (iv) $(A \oplus B) \cap C = (A \cap C) \oplus (B \cap C)$.

8 Prove the identities (i) $A \times (B \cap C) = (A \times B) \cap (A \times C)$ and (ii) $A \times (B \cup C) = (A \times B) \cup (A \times C)$.

9 Prove that $(A \times B) \cap (B \times A) = (A \cap B) \times (A \cap B)$, for any sets A, B. Is the same true with $\cap$ replaced by $\cup$?

10 Prove the identity $(A \cup B) \times (C \cup D) = (A \times C) \cup (A \times D) \cup (B \times C) \cup (B \times D)$. (This compares with the algebraic identity $(a + b)(c + d) = ac + ad + bc + bd$.)

11 Mark on a co-ordinate plane the regions which represent the following subsets of $R \times R$ (see *Example 23*, p. 11):

(i) $\{(x, y) | x = 0\}$;   (ii) $\{(x, y) | x > y\}$;   (iii) $\{(x, y) | x^2 + y^2 = 1\}$;
(iv) $\{(x, y) | x^2 + y^2 < 1\}$; (v) $\{(x, y) | 0 \leq x \leq 1, 0 \leq y \leq 1\}$.

12 For each positive integer n, let $Q_n = \{x \in R | 0 \leq x < (n+1)/n\}$. Let $\mathscr{S}$ be the set of all these sets Q_n, for $n = 1, 2, \ldots$. Show that $\cap \mathscr{S} = \{x \in R | 0 \leq x \leq 1\}$. Find also $\cup \mathscr{S}$.

13 Prove that the number of subsets (including $\varnothing$) of a finite set of order n is 2^n. (Use induction on n. See also *Example 35*, p. 19.)

2 Equivalence relations

2.1 Relations on a set

In many sets there occur relations which hold between certain pairs of elements. A relation is usually described by a *statement* which involves an arbitrary pair of elements of the set. For example the statement 'x is the mother of y' describes a relation on the set P of all living people; for certain pairs (x, y) of people the statement is true, and for all other pairs it is false. Similarly '$x > y$' (meaning 'x is greater than y') describes a relation on the set R of all real numbers.

As a general notation we shall often use $\sim$ for a relation on a set A.

DEFINITION. $\sim$ is a *relation*† *on the set A*, if, for each pair (x,y) of elements of A, the statement '$x \sim y$' has a meaning, i.e. it is either true or false for that particular pair. We write simply $x \sim y$, to mean that '$x \sim y$' is *true*, for a given pair (x,y).

For example, taking $A = P$, we could use '$x \sim y$' as an abbreviation for 'x is the mother of y', and then $\sim$ is a relation on P. Similarly $>$ is a relation on R, because for every pair (x, y) of real numbers, the statement '$x > y$' has a meaning, i.e. it is either true or false for that particular pair. When we write simply $x > y$, with no quotation marks, then of course we mean that the statement '$x > y$' is *true*.

A very familiar relation, applicable to any set A, is the relation of *equality*, denoted by $=$, so that '$x = y$' means, as usual, 'x is equal to y'.

---

†Often called a *binary relation*, because it relates pairs of elements of A.

2.2 Equivalence relations

Now we shall describe some special kinds of relation. Let A be any set, and let $\underset{\rho}{\sim}$ be a relation on A.

DEFINITION 1. $\sim$ is a *reflexive* relation if it satisfies the condition

E1. If x is any element of A, then $x \sim x$.

DEFINITION 2. $\sim$ is a *symmetric* relation if it satisfies

E2. If x, y are any elements of A such that $x \sim y$, then also $y \sim x$.

DEFINITION 3. $\sim$ is a *transitive* relation if it satisfies

E3. If x, y, z are any elements of A such that $x \sim y$ and $y \sim z$, then also $x \sim z$.

DEFINITION 4. $\sim$ is an *equivalence relation*† if it is reflexive, symmetric and transitive, i.e. if it satisfies all three conditions **E1**, **E2**, **E3**.

Example 29. The relation of equality is the simplest equivalence relation. If we take $x \sim y$ to mean $x = y$, clearly **E1**, **E2**, **E3** are satisified.

Example 30. Define a relation $\sim$ on the set P of all people, by taking '$x \sim y$' to mean 'x and y have the same age' (meaning, to be precise, that their ages on their last birthdays were the same). This is an equivalence relation on P, as we see at once by checking that **E1**, **E2**, **E3** are satisfied.

Example 31. If we define $\sim$ to be the relation 'is the mother of' on the set P, we find that this is neither reflexive, symmetric nor transitive. Therefore it is not an equivalence relation on P.

Example 32. The relation $>$ on R is *transitive*, for if x, y, z are any numbers such that $x > y$ and $y > z$, clearly also $x > z$. But it is not reflexive, since '$x > x$' is never true, and it is not symmetric (take for example $x = 2$, $y = 1$ to show that **E2** is not satisfied). So $>$ is not an equivalence relation.

---

†The symbols $\simeq$, $\cong$, $\equiv$ are often used to denote equivalence relations.

2.3 Partitions

In the next section we shall prove that any equivalence relation on a set A determines a *partition* of A. So we must interrupt our discussion of relations, to define this new idea.

DEFINITION. A *partition* of a set A is a set $\mathscr{P}$ of non-empty subsets of A, such that each element x of A belongs to one and only one member of $\mathscr{P}$.

Figure 7 illustrates the way a partition 'divides up' the set A. In this case $\mathscr{P}$ is the set with five elements X_1, X_2, X_3, X_4, X_5, which are subsets of A, such that each element of A lies in one and only one of them.

Example 33. Let Z^+ be the set of all positive integers, Z^- the set of all negative integers, and Z^0 the one-element set $\{0\}$. Then $\mathscr{P} = \{Z^+, Z^-, Z^0\}$ is a partition of the set Z of all integers, because every integer z belongs to one and only one of Z^+, Z^- and Z^0.

Example 34. We get a partition of a set A as soon as we *classify* the elements of A in such a way that each element x of A falls into one and only one 'class', i.e. subset. For example, we might classify people according to their age on their last birthday, and let X_n denote the set of all people of age n; then the set of all these sets X_n is a partition of the set P of all people.

Partitions of a finite set. Suppose that $\mathscr{P}$ is a partition of a finite set A into n subsets, $\mathscr{P} = \{X_1, \ldots, X_n\}$, say. By counting up all the elements in these sets $X_1, \ldots, X_n$ we must count each element of A, and of course we have counted no element twice, because we

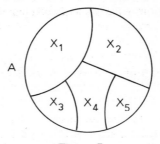

Figure 7

know that no element x of A belongs to two of the sets X_1. This gives the simple formula

$$|A| = |X_1| + \ldots + |X_n|,$$

which is often very useful.

Example 35. Let $B = \mathcal{B}(X)$ be the set of all subsets of a set X which has n elements. For each integer r in the range $0 \leq r \leq n$, let Y_r be the set of all subsets of X which have order r. For example, Y_0 contains only the empty set $\varnothing$, Y_1 is the set of all the one-element subsets of X, and so on. Now $\{Y_0, Y_1, \ldots, Y_n\}$ is a partition of B, because every element of B belongs to exactly one of these sets, hence $|B| = |Y_0| + |Y_1| + \ldots + |Y_n|$. From elementary algebra, $|Y_r|$, the number of ways of choosing a set of r elements from the n elements of X, is the 'binomial coefficient'

$$\binom{n}{r} = \frac{n!}{r!\,(n-r)!}$$

(often written ^nC_r), i.e. it is the coefficient of t^r in the expansion of $(1+t)^n$ in powers of t. Thus

$$|Y_0| + |Y_1|t + |Y_2|t^2 + \ldots + |Y_n|t^n = (1+t)^n,$$

and putting $t = 1$ we find

$$|B| = |\mathcal{B}(X)| = |Y_0| + |Y_1| + \ldots + |Y_n| = 2^n.$$

See also Example 61, p. 39.

2.4 Equivalence classes

DEFINITION. Let A be any set, and $\sim$ any equivalence relation on A. For each fixed element x of A, the *equivalence class* E_x of x (with respect to the relation $\sim$) is the set of all elements y of A such that $x \sim y$, i.e.

$$E_x = \{y \in A \,|\, x \sim y\}.$$

Thus E_x is a subset of A. Since $x \sim x$ by **E1**, we see that x itself belongs to E_x.

Example 36. Let $\sim$ be the equivalence relation on P defined in Example 30, p. 17. If x is a given person, then E_x (with respect to this relation) is the set of all people y who have the same age as x; if the age of x

is n years, then E_x is the set X_n of Example 34. Notice that E_y also is the same set, for any other person y of the same age, in fact, $E_x = E_y$ if and only if $x \sim y$. The equivalence classes E_x, for all $x \in P$, are the sets X_n of Example 34, and these form a partition of P. The reader may find this example useful in understanding the theorem which follows.

FUNDAMENTAL THEOREM ON EQUIVALENCE RELATIONS

If $\sim$ is an equivalence relation on a set A, then the set $\mathscr{P} = \{E_x | x \in A\}$ of all equivalence classes with respect to $\sim$ is a partition of A. Moreover if x, y are any elements of A, then $E_x = E_y$ if and only if $x \sim y$.

We shall prove this in three steps, in the first two of which are proved two 'lemmas', i.e. auxiliary facts needed in the final step.

(i) LEMMA. *If $x \sim y$, then $E_x = E_y$.*

Proof. Suppose that $x \sim y$, and that z is any element of E_y, which means that $y \sim z$. Now we have $x \sim y$ and $y \sim z$, therefore $x \sim z$ by **E3**. Hence $z \in E_x$. Thus every element of E_y is an element of E_x. But the symmetric property **E2** shows that also $y \sim x$, and the same argument shows that every element of E_x is an element of E_y. Therefore $E_x = E_y$.

(ii) LEMMA. *If $E_x \cap E_y \neq \varnothing$, i.e. if E_x, E_y have any element in common, then $x \sim y$, and hence also $E_x = E_y$.*

Proof. Suppose there is an element z in $E_x \cap E_y$. The $z \in E_x$ and $z \in E_y$, which means $x \sim z$ and $y \sim z$. From $y \sim z$ follows $z \sim y$ by **E2**. Now we have $x \sim z$ and $z \sim y$, therefore $x \sim y$ by **E3**. From lemma (i) it follows that also $E_x = E_y$.

(iii) *Proof of the theorem.* To show that $\mathscr{P}$ is a partition of A, we have to show that each element x of A belongs to one and only one equivalence class. Now we saw on p. 19 that x belongs to the equivalence class E_x. If x also belongs to an equivalence class E_y, then E_x and E_y have the element x in common, so that $E_x = E_y$ by lemma (ii).

Finally we must prove that $E_x = E_y$ if and only if $x \sim y$. If $x \sim y$ then $E_x = E_y$ by lemma (i). If $E_x = E_y$ then $x \sim y$ by lemma (ii). This completes the proof of the theorem.

Remark

If E is any one of the equivalence classes with respect to $\sim$, there are in general many different ways of representing E. In fact, *if x is any element of E, then $E = E_x$*. For, as we have seen, there is only one equivalence class which contains x. But we know $x \in E_x$ (because $x \sim x$) and also $x \in E$. Hence $E = E_x$. So if E has elements $x, y, z, \ldots$, then $E = E_x = E_y = E_z = \ldots$.

Example 37. Suppose we start with any partition $\mathscr{D}$ of a set A. Then we may define a relation $\sim$ on A by the rule, '$x \sim y$' shall mean 'x and y belong to the same member of $\mathscr{D}$'. It is easy to verify that this is an equivalence relation on A, and that the equivalence classes with respect to $\sim$ are precisely the sets which are members of $\mathscr{D}$. This provides a converse to the theorem above, and shows that *equivalence relations* and *partitions*, of a given set A, are practically interchangeable.

2.5 Congruence of integers

For the rest of this chapter we shall discuss in more detail a particular example of an equivalence relation, which is very useful in the theory of numbers (i.e. of integers), and has given rise to many important ideas in algebra.

Let m be a positive integer; we shall keep m fixed for the rest of the chapter. If x, y are any integers we make the following definition.

DEFINITION. x is congruent to y modulo m, if $x - y$ is a multiple of m, i.e. if there exists an integer k such that $y = x + km$. We denote this†

$$x \equiv y \bmod m.$$

It is easy to verify that, for fixed m, this relation of 'congruence mod m' is an *equivalence relation on the set* $\mathbf{Z}$ *of all integers*.

E1 holds, because for any $x \in \mathbf{Z}$, $x - x = 0$ is a multiple of m, hence $x \equiv x \bmod m$. **E2**. Suppose $x \equiv y \bmod m$, i.e. $x - y$ is a multiple of m.

---

† 'Modulo' means 'with modulus (or measure)'. The idea of congruence of integers was introduced by Gauss (1777–1855). We use 'mod m' as an abbreviation for 'modulo m'.

Then the same is true of $y-x=-(x-y)$, so $y\equiv x$ mod m **E3**.
Suppose $x\equiv y$ and $y\equiv z$ mod m, i.e. $x-y$ and $y-z$ are multiples of
m. Then the same is true of $x-z=(x-y)+(y-z)$, and so $x\equiv z$
mod m.

Example 38. $13\equiv 3$ mod 5, $-10\equiv 8$ mod 6, $-3\equiv -5$ mod 2. Any two
integers x, y are congruent mod 1. To say that $x\equiv 0$ mod m is the same as
saying x is a multiple of m. If $x\equiv y$ mod m, and if n is a factor (divisor) of m,
then $x\equiv y$ mod n.

Congruence classes (residue classes)

Let x be a given integer. The equivalence class E_x with respect to
the relation of congruence mod m, is called the *congruence class*, or
residue class of x mod m. E_x is the set of all integers y such that $x\equiv y$
mod m, i.e. $E_x=\{x+km|k\in Z\}=$

$$\{ \ldots, -2m+x, -m+x, x, x+m, x+2m, \ldots \}.$$

Let r be the smallest integer r in E_x which is ≥ 0; r is called the
residue of x mod m. r must be one of the integers $0, 1, \ldots, m-1$,
because if $r\geq m$, then $r-m$ is also in E_x and is ≥ 0, but is smaller
than r. Also $E_x=E_r$, by the fundamental theorem on p. 20, because
$x\equiv r$ mod m. Thus each class E_x is equal to one of the m classes E_0,
$E_1, \ldots, E_{m-1}$; moreover these are distinct sets, because no two of
$0, 1, \ldots m-1$ are congruent mod m. To summarize, *there are
exactly m different congruence classes mod m, namely E_0, $E_1, \ldots$,
E_{m-1}, and these form a partition of Z. The class E_r consists of all
integers which have a given residue r mod m.*

Example 39. If x is positive, then the residue r is the same as the *remainder
on division by m*. For when we divide x by m we subtract as large a multiple
of m as we can, say qm, from x, without making the difference $x-qm$
negative. But $x-qm$ belongs to E_x, for all $q\in Z$, so the remainder
$r=x-qm$ is the smallest element of E_x which is ≥ 0. For example, on
dividing 29 by 8 we get 'quotient' $q=3$, so $5=29-3.8$ is the residue of 29
mod 8.

Example 40. Taking $m=4$, the four congruence classes are

$$E_0 = \{\ldots, -8, -4, 0, 4, 8, 12, \ldots\},$$
$$E_1 = \{\ldots, -7, -3, 1, 5, 9, 13, \ldots\},$$
$$E_2 = \{\ldots, -6, -2, 2, 6, 10, 14, \ldots\}$$

and

$$E_3 = \{\ldots, -5, -1, 3, 7, 11, 15, \ldots\}.$$

It is easy to see that every integer is in one and only one of these.

2.6   Algebra of congruences

Congruence relations have a property which distinguishes them from other equivalence relations on Z, namely that two congruences mod m can be added, subtracted or multiplied like ordinary equations. By this we mean the following: *If x, x', y, y' are integers such that $x \equiv x'$ mod m and $y \equiv y'$ mod m, then* (i) $x + y \equiv x' + y'$ *mod m,* (ii) $x - y \equiv x' - y'$ *mod m and* (iii) $xy \equiv x'y'$ *mod m.*

Proof. We are given that $x - x'$ and $y - y'$ are both multiples of m. Therefore $(x + y) - (x' + y') = (x - x') + (y - y')$ is a multiple of m, which proves (i). The proof of (ii) is just as easy. Finally $xy - x'y' = x(y - y') + (x - x')y'$, showing that this too is a multiple of m, and this proves (iii).

For an account on many interesting applications of congruences, the reader is referred to the book *The Higher Arithmetic*, by H. Davenport (Hutchinson's University Library).

Example 41. Tests for divisibility. Applying (iii) repeatedly we can see that if $x \equiv y$ mod m, then $x^n \equiv y^n$ mod m, for any positive integer n. For example $10 \equiv 1$ mod 9, hence $10^n \equiv 1^n = 1$ mod 9, for any n. Now let $x = b_1 b_{r-1} \ldots b_1 b_0$ be any positive integer, written in ordinary decimal notation. Then $x = b_0 + 10b_1 + \ldots + 10^r b_r$, and by the rules we have proved, $x \equiv b_0 + 1 \cdot b_1 + \ldots + 1 \cdot b_r$ mod 9, i.e. x has the same residue mod 9 as $b_0 + b_1 + \ldots + b_r$, the sum of the digits of x; in particular, x is divisible by 9 if and only if this sum is divisible by 9. A similar test for divisibility by 3 works because $10 \equiv 1$ mod 3. See also Exercise 9, p. 25.

Example 42. The *binomial theorem* for a positive exponent p gives

$$(1 + t)^p = 1 + pt + \frac{p(p-1)}{1 \cdot 2} t^2 + \ldots + t^p \tag{2.1}$$

t being a variable. The coefficient of $t^r (0 \le r \le p)$ is

$$\binom{p}{r} = \frac{p(p-1)\dots(p-r+1)}{1.2\dots r}. \tag{2.2}$$

Now if p is a *prime* number and if r is one of $1, 2, \dots, p-1$, then none of the factors $1, 2, \dots, r$ in the denominator of equation (2.2) cancels the factor p in the numerator, hence $\binom{p}{r}$ is a multiple of p, i.e.

$$\binom{p}{r} \equiv 0 \bmod p.$$

Putting this in equation (2.1) we get

$$(1+t)^p \equiv 1 + t^p \bmod p. \tag{2.3}$$

For example, $(1+t)^5 = 1 + 5t + 10t^2 + 10t^3 + 5t^4 + t^5 \equiv 1 + t^5 \bmod 5$. Now raise both sides of equation (2.3) to the pth power. We get $(1+t)^{p^2}$ on the left, and on the right $(1+t^p)^p$, which by another application of equation (2.3) (putting t^p for t in equation (2.3)) is $\equiv 1 + t^{p^2} \bmod p$. Going on like this we find

$$(1+t)^{p^a} \equiv 1 + t^{p^a} \bmod p, \tag{2.4}$$

for any prime p, and any positive integer a. We use this now to prove a fact which will be needed later (Example 125, p. 76): *If p is a prime and if k, a are any positive integers, then*

$$\binom{kp^a}{p^a} \equiv k \bmod p. \tag{2.5}$$

Proof. Write

$$h = \binom{kp^a}{p^a}$$

for short. This is the coefficient of t^{p^a} in $(1+t)^{kp^a}$. We can write

$$(1+t)^{kp^a} = [(1+t)^{p^a}]^k,$$

so by equation (2.4),

$$(1+t)^{kp^a} \equiv (1+t^{p^a})^k \bmod p.$$

But if we use the binomial theorem to expand $(1+t^{p^a})^k$, we find that the coefficient of t^{p^a} is k. Therefore $h \equiv k \bmod p$, which proves equation (2.5).

Exercises for Chapter 2

1 Find whether the following relations are reflexive, symmetric or transitive. Which are equivalence relations? (i) $A = R$ and '$x \sim y$' means '$x - y$ is an integer'. (ii) $A = R$ and '$x \sim y$' means '$x + y$ is an integer'. (iii) $A = P$ and 'x and y have an ancestor in common'. (iv) A is the set of all positive integers, and '$x \sim y$' means 'x divides y' (i.e. y/x is an integer).

2 For each integer n let $X_n = \{x \in R | n \leq x < n+1\}$. Prove that $P = \{X_n | n \in Z\}$ is a partition of R.

3 For any real number x, let $[x]$ ('integral part of x') denote the largest integer $\leq x$, e.g. $[\sqrt{5}] = 2$, $[-\frac{1}{2}] = -1$, $[3] = 3$. Define a relation $\sim$ on R by letting '$x \sim y$' mean '$[x] = [y]$'. Prove that $\sim$ is an equivalence relation and that its equivalence classes are the sets X_n of Exercise 2.

4 Prove that the relation $\sim$ of Example 37, p. 21, is an equivalence relation, and that its equivalence classes are the members of $\mathscr{D}$.

5 Write down the different congruence classes (i) mod 2; (ii) mod 3.

6 Find the residue of 2^{512} mod 5.

7 If p is a prime prove that $xy \equiv 0 \bmod p$ implies that $x \equiv 0 \bmod p$ or $y \equiv 0 \bmod p$ (or both). Prove that this fails if p is not prime.

8 Let A be the congruence class of 1 mod 3, and B the congruence class of -1 mod 4. Prove that $A \cap B$ is a congruence class mod 12.

9 Let $x = b_r b_{r-1} \ldots b_1 b_0$ be a positive integer written in decimal notation. (i) Show that x is divisible by 11 if and only if $b_0 - b_1 + b_2 - \ldots$ is divisible by 11. (ii) Show that x is divisible by 12 if and only if $b_0 - 2b_1 + 4(b_2 + b_3 + \ldots)$ is divisible by 12.

3 Maps

3.1 Maps

The idea of a *map* (or *mapping*) of one set into another is one of the most fruitful of modern mathematics. Like the other two 'abstract' notions which we have already met, namely those of *set* and *relation*, it is an elementary idea, but it gives us a new power to express accurately some fundamental mathematical situations. In particular, maps give us a way to describe connections between the elements of different sets, or of the same set.

DEFINITION. Let A, B be any sets (which may be equal). Then a *map* θ *of A into B* is determined when there is a rule or formula which assigns, to each element x of A, an element of B, called the *image of x under* θ. This element of B is usually written in one of two standard notations

$$\theta(x), \qquad \text{or} \qquad x\theta.$$

In this book we shall adopt the first notation.

We can imagine θ as a kind of agent which 'transforms' or 'maps' or 'projects' each element x of A to its image element $\theta(x)$ in B (see Figure 8).

The words *transformation, projection, operation, substitution* are often used for maps of various kinds. We may also think of θ as a general kind of 'function', with arguments x in A and values $\theta(x)$ in B. We shall often use Greek letters, such as θ, ϕ, ψ, α, β, etc. for maps.

Example 43. We can define a map θ from the set P of all people into the set Z of all integers, by the rule: if $x \in P$, then $\theta(x)$ is the age of x (i.e. $\theta(x)$ is his age, in years, on his last birthday).

Example 44. There are in general many different maps of a given set into another. For example we may define a second map ϕ of P into Z by the rule: if $x \in P$, let $\phi(x)$ be the height of x, measured in millimetres to the nearest millimetre.

Example 45. Ordinary *functions* provide examples of maps. For example, let θ be the map of R (the set of all real numbers) into itself defined by the rule: if $x \in R$, let $\theta(x) = e^x$. This is a map of R into R.

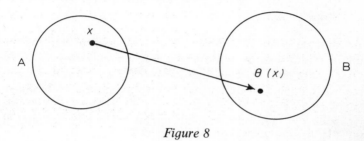

Figure 8

3.2 Equality of maps

It is convenient to use one of the notations

$$\theta: A \longrightarrow B \qquad \text{or} \qquad A \overset{\theta}{\longrightarrow} B$$

to indicate that θ is a map of A into B. The set A is called the *domain* of θ, and B is called the *range* of θ. Notice that we do not assume that every element of the range is the image of some element of the domain (see section 3.3).

DEFINITION. Two maps θ and ϕ are equal (and we write $\theta = \phi$) if and only if (i) θ, ϕ have the same domain, (ii) θ, ϕ have the same range, and (iii) $\theta(x) = \phi(x)$ for each element x of the domain.

Example 46. Our definition requires that *all* of the conditions (i), (ii), (iii) be satisfied. Consider the map $\theta: R \to R$ of Example 45. We know from elementary calculus that e^x is always *positive*, for all $x \in R$. So we could define a map $\phi: R \to R^+$, where R^+ is the set of all positive real numbers, by the rule $\phi(x) = e^x$. The only difference between θ and ϕ is that they have different ranges, i.e. (i) and (iii) are satisfied, but not (ii). Therefore θ and ϕ are *not* equal, according to our definition.†

Maps of finite sets

If $A = \{a_1, \ldots, a_m\}$ is a finite set, and B is any set, we can represent a map $\theta: A \to B$ by the symbol

$$\theta = \begin{pmatrix} a_1 & & a_m \\ \theta(a_1) & \cdots & \theta(a_m) \end{pmatrix}.$$

For example if $A = \{1, 2, 3\}$ and $B = \{a, b\}$, then $\theta = \begin{pmatrix} 123 \\ aba \end{pmatrix}$ is the map of A into B defined by $\theta(1) = a$, $\theta(2) = b$, $\theta(3) = a$.

If A, B are both finite, with orders m, n respectively, then there are n^m different maps θ of A into B. For we can choose, as the image $\theta(x)$ of each element x of A, any one of the n elements of B; this gives $n \times n \times \ldots \times n = n^m$ possibilities. Thus the $2^3 = 8$ mappings of $A = \{1, 2, 3\}$ into $B = \{a, b\}$ are

$$\begin{pmatrix} 123 \\ aaa \end{pmatrix}, \begin{pmatrix} 123 \\ aab \end{pmatrix}, \begin{pmatrix} 123 \\ aba \end{pmatrix}, \begin{pmatrix} 123 \\ abb \end{pmatrix}, \begin{pmatrix} 123 \\ baa \end{pmatrix}, \begin{pmatrix} 123 \\ bab \end{pmatrix}, \begin{pmatrix} 123 \\ bba \end{pmatrix}, \begin{pmatrix} 123 \\ bbb \end{pmatrix}.$$

3.3 Injective, surjective, bijective maps. Inverse maps

The definition of a map does *not* require that two distinct elements x, x' of the domain should have distinct images. In Example 43, we have $\theta(x) = \theta(x')$ whenever x, x' are two people of the same age. Also we do *not* require that every element y of the range should be the image of some element x of the domain; in the same example, if we take $y = -3$ or $y = 1000$, then there is no x in P such that $\theta(x) = y$. However, it is useful to have special names for those maps which do satisfy these special conditions.

DEFINITION 1. A map $\theta: A \to B$ is *injective*‡ if whenever x, x' are distinct elements of A, then $\theta(x)$, $\theta(x')$ are distinct elements of B.

---

†This is a matter of convention; some authors would say that θ, ϕ are equal if (i) and (iii) hold. However, this makes it difficult to define inverse maps; see Example 48.
‡Sometimes the word 'one-to-one' is used instead of 'injective'.

Equivalently, θ is injective if

$$\theta(x) = \theta(x') \Rightarrow x = x',$$

for every pair of elements x, x' of A. An injective map is sometimes called an *injection*.

DEFINITION 2. A map $\theta: A \to B$ is *surjective*† if for each element y of B there is at least one element x of A such that $\theta(x) = y$. A surjective map is sometimes called a *surjection*.

DEFINITION 3. A map $\theta: A \to B$ is *bijective* if it is both injective and surjective. A bijective map is sometimes called a *bijection*.

Bijective maps are particularly important. If $\theta: A \to B$ is bijective, it means that, for each $y \in B$, there is a *unique* element $x \in A$ such that $\theta(x) = y$ (the *uniqueness* of x comes from the fact that θ is injective, while the *existence* of x comes from the fact that θ is surjective). Therefore we can define a map $\theta^{-1}: B \to A$, called the *inverse of* θ, by the rule: if $y \in B$, let $\theta^{-1}(y) = x$, where x is the unique element of A such that $\theta(x) = y$. Notice that θ^{-1} is defined only when θ is bijective. It is easy to see that θ^{-1} is itself bijective, and that $(\theta^{-1})^{-1} = \theta$.

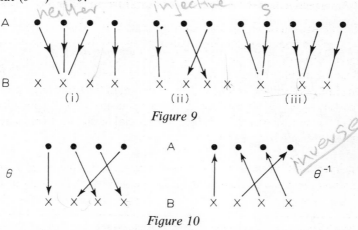

Figure 9

Figure 10

Example 47. Three maps $\theta: A \rightarrow B$ are shown in Figure 9 (in each case, the dots represent elements of A, and the crosses elements of B). In (i) θ is neither injective nor surjective, (ii) is injective but not surjective, (iii) is surjective but not injective. Figure 10 represents a bijective map and its inverse. All the eight maps at the end of section 3.2 are surjective, except the first and last. None is injective.

Example 48. The map $\theta: R \rightarrow R$ of Example 45 is injective, because if $e^x = e^{x'}$ then $x = x'$. It is not surjective, because if y is any negative number, or zero, there is no real x such that $e^x = y$. But the map $\phi: R \rightarrow R^+$ of Example 46 is both injective and surjective, i.e. ϕ is bijective. Therefore there is an inverse map $\phi^{-1}: R^+ \rightarrow R$. In fact ϕ^{-1} is given by the rule: if $y \in R^+$, then $\phi^{-1}(y) = \log y$.

3.4 Product of maps

Suppose that A, B, C are any sets, and that we have maps $\theta: A \rightarrow B$ and $\phi: B \rightarrow C$. Then θ maps each element x of A to its image $\theta(x)$, which is an element of B. If we now apply ϕ to $\theta(x)$, we get the element $\phi[\theta(x)]$ of C. So the result of applying first θ and then ϕ, is a new map from A into C. This map is called the *product* (sometimes called the *composite*) of θ and ϕ, and is denoted $\phi\theta$. So we have the following.

DEFINITION. If $\theta: A \rightarrow B$ and $\phi: B \rightarrow C$, then $\phi\theta: A \rightarrow C$ is the map defined by the rule: if $x \in A$, then $(\phi\theta)(x) = \phi[\theta(x)]$. Note that $\phi\theta$ results from applying *first* θ and *then* ϕ.

Example 49. Take $A = \{1, 2, 3\}$, $B = \{a, b\}$, $C = \{u, v, w\}$, and let

$$\theta = \begin{pmatrix} 123 \\ aba \end{pmatrix}, \qquad \phi = \begin{pmatrix} ab \\ wv \end{pmatrix}.$$

Then

$$\phi\theta = \begin{pmatrix} 123 \\ wvw \end{pmatrix};$$

because, for example, θ takes 1 to a, then ϕ takes a to w, so $\phi\theta$ takes 1 to w. Similarly $\phi\theta$ takes 2 to v, and 3 to w.

Example 50. Take $A = B = C = \{1, 2, 3\}$, and consider the two maps

$$\alpha = \begin{pmatrix} 123 \\ 231 \end{pmatrix}, \qquad \rho = \begin{pmatrix} 123 \\ 132 \end{pmatrix}.$$

Then

$$\alpha\rho = \begin{pmatrix} 123 \\ 213 \end{pmatrix} \qquad \text{and} \qquad \rho\alpha = \begin{pmatrix} 123 \\ 321 \end{pmatrix},$$

which are both maps of A into A, but $\alpha\rho \neq \rho\alpha$.

This 'multiplication' of maps is very different from ordinary multiplication of numbers. In the first place, the product $\phi\theta$ exists only if the range of θ is the same as the domain of ϕ. For example, there is no product $\theta\phi$ in Example 49. And Example 50 shows that even if both products $\phi\theta$ and $\theta\phi$ exist, they need not be equal. If it happens that $\phi\theta = \theta\phi$, we say that the maps θ and ϕ *commute.*

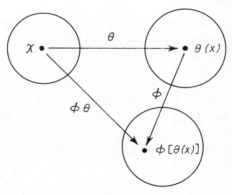

Figure 11

Product of several maps

Suppose that A, B, C, D are any sets, and that we have maps $\theta: A \to B$, $\phi: B \to C$, $\psi: C \to D$. Then we define $\psi\phi\theta: A \to D$ to be the map which takes an element x of A to $\psi\{\phi[\theta(x)]\}$, i.e. we follow the route θ, ϕ, ψ in Figure 12.

But we also see from Figure 12, that $\psi\phi\theta$ is the same as the

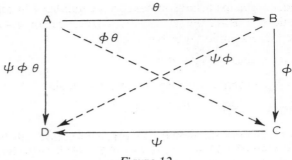

Figure 12

product of $\phi\theta$ and ψ, or equally, of θ and $\psi\phi$. This gives the *associative law for maps*,

$$\psi(\phi\theta) = (\psi\phi)\theta.$$

Of course this law applies only if the products involved are all defined, as in the case we have shown. We can extend it to products of n maps $(n > 3)$. If A_i $(i = 1, 2, \ldots, n+1)$ are sets, and $\theta_i : A_i \to A_{i+1}$ $(i = 1, 2, \ldots, n)$ are maps, then

$$\theta_n \ldots \theta_2\theta_1 : A_1 \to A_{n+1}$$

is defined to be the map obtained by applying $\theta_1, \theta_2, \ldots, \theta_n$ in turn. This product is the same, however it is 'bracketed'. For example $(\theta_4\theta_3)(\theta_2\theta_1)$ equals $\theta_4[(\theta_3\theta_2)\theta_1]$, because both are equal to $\theta_4\theta_3\theta_2\theta_1$.

Example 51. Powers. If θ is a map of A into itself, then it is possible to define the *powers* $\theta, \theta^2, \theta^3, \ldots$ of θ; they are all maps of A into itself, θ^n being the map obtained by applying θ n times. For example, $\theta^2 = \theta\theta$, $\theta^3 = \theta\theta\theta$. Any two powers of θ commute, because if m, n are any positive integers, then $\theta^m\theta^n$ and $\theta^n\theta^m$ are both equal to θ^{m+n}, hence $\theta^m\theta^n = \theta^n\theta^m$.

3.5 Identity maps

DEFINITION. If A is any set, the *identity map on A* is the map $\iota_A : A \to A$ defined by the rule: if $x \in A$, then $\iota_A(x) = x$.
Clearly ι_A is bijective, and $\iota_A^{-1} = \iota_A$. With respect to the product of

maps, identity maps behave rather like the number 1 in ordinary arithmetic. If A, B are any sets and $\theta: A \to B$ is any map of A into B, we have (see Example 52, below)

$$\theta \iota_A = \theta \qquad \text{and} \qquad \iota_B \theta = \theta.$$

If $\theta: A \to B$ is bijective, then (see Example 52 again)

$$\theta^{-1}\theta = \iota_A \qquad \text{and} \qquad \theta\theta^{-1} = \iota_B.$$

Example 52. To prove the formula $\theta \iota_A = \theta$, notice first that both $\theta \iota_A$ and θ are maps of A into B. Then take any $x \in A$. We have $(\theta \iota_A)(x) = \theta[\iota_A(x)] = \theta(x)$. This proves that $\theta \iota_A = \theta$. To prove $\theta^{-1}\theta = \iota_A$ (it being now assumed that θ is bijective) notice first that $\theta^{-1}\theta$ and ι_A are both maps of A into A. Then take any $x \in A$. If $\theta(x) = y$, then by the definition of θ^{-1}, $\theta^{-1}(y) = x$. Therefore $(\theta^{-1}\theta)(x) = \theta^{-1}[\theta(x)] = \theta^{-1}(y) = x = \iota_A(x)$. Referring to the definition of equality of maps (p. 27), we see that $\theta^{-1}\theta = \iota_A$. The formulae $\iota_B \theta = \theta$ and $\theta\theta^{-1} = \iota_B$ are proved similarly.

3.6 Products of bijective maps

Let A, B, C be any sets, and suppose that $\theta: A \to B$, $\phi: B \to C$ are given maps. We prove next

(i) *If θ, ϕ are both injective, then $\phi\theta$ is injective. If θ, ϕ are both surjective, then $\phi\theta$ is surjective.*

Proof. Suppose first that θ, ϕ are both injective, and that x, x' are any two distinct elements of A. Then $\theta(x)$, $\theta(x')$ are distinct because θ is injective, hence also $\phi[\theta(x)]$, $\phi[\theta(x')]$ are distinct because ϕ is injective. Thus $(\phi\theta)(x)$, $(\phi\theta)(x')$ are distinct, whenever x, x' are distinct. This proves that $\phi\theta$ is injective.

Now suppose that θ, ϕ are both surjective (but neither is necessarily injective). Let z be any element of C. Because ϕ is surjective, there is some y in B such that $\phi(y) = z$. And because θ is surjective there is some x in A such that $\theta(x) = y$. Hence $(\phi\theta)(x) = \phi[\theta(x)] = \phi(y) = z$. This proves that for any z in C there is some x in A such that $(\phi\theta)(x) = z$, hence $\phi\theta$ is surjective. We have now proved both statements in (i). If we put these two statements together, we get

(ii) *If θ, ϕ are both bijective, then $\phi\theta$ is bijective.*

There is also a partial converse to each of the statements in (i). Suppose first that $\theta: A \to B$ and $\phi: B \to C$ are such that $\phi\theta$ is injective. Then θ *must be injective*. For otherwise, there would exist distinct x, x' in A such that $\theta(x) = \theta(x')$. Applying ϕ to both sides of this equation gives $(\phi\theta)(x) = (\phi\theta)(x')$. But this contradicts our assumption that $\phi\theta$ is injective. In other words, we must have θ injective, if $\phi\theta$ is injective.

Similarly if θ, ϕ are such that $\phi\theta$ is surjective, then ϕ *must be surjective*. For take any z in C. Since $\phi\theta$ is surjective, there is some x in A such that $(\phi\theta)(x) = z$, i.e. $\phi[\theta(x)] = z$. Therefore we can always find y in B such that $\phi(y) = z$, namely $y = \theta(x)$. This proves that ϕ is surjective.

With the help of these remarks we shall prove the following lemma, which is often used to prove that a given map is bijective.

(iii) LEMMA. *Let $\theta: A \to B$ be a map, and suppose that there exists a map $\theta': B \to A$ which satisfies $\theta'\theta = \iota_A$ and $\theta\theta' = \iota_B$. Then θ is bijective, and $\theta' = \theta^{-1}$.*

Proof. Since $\theta'\theta = \iota_A$, which is injective, then θ is injective. And since $\theta\theta' = \iota_B$, which is surjective, then θ is surjective. Hence θ is bijective. To prove $\theta' = \theta^{-1}$, calculate the product $\theta'\theta\theta^{-1}$ in two ways. First $\theta'\theta\theta^{-1} = \theta'(\theta\theta^{-1}) = \theta'\iota_B = \theta'$, by the results in the last section. But also $\theta'\theta\theta^{-1} = (\theta'\theta)\theta = \iota_A\theta^{-1}$ (since we are given $\theta'\theta = \iota_A$) $= \theta^{-1}$. Hence $\theta' = \theta^{-1}$.

Example 53. If θ, ϕ are bijections, then $(\phi\theta)^{-1} = \theta^{-1}\phi^{-1}$. This is easy to prove directly, but we shall prove it here as an application of lemma (iii). Put $\phi\theta = \alpha$ and $\theta^{-1}\phi^{-1} = \beta$. Then $\alpha: A \to C$ and $\beta: C \to A$ satisfy $\beta\alpha = \iota_A$ and $\alpha\beta = \iota_C$ (for example $\beta\alpha = \theta^{-1}\phi^{-1}\phi\theta = \theta^{-1}\iota_B\theta = \theta^{-1}\theta = \iota_A$). By the lemma, $\alpha = \phi\theta$ is bijective and $\beta = \alpha^{-1}$, i.e. $\theta^{-1}\phi^{-1} = (\phi\theta)^{-1}$.

3.7 Permutations

DEFINITION. If A is a set, then a *permutation* is a bijection of A into itself, i.e. a permutation of A is a map $\theta: A \to A$ which is both injective and surjective.

For example ι_A is a permutation of A. If θ, ϕ are permutations of A, then θ^{-1} and $\phi\theta$ are also permutations of A. We shall need these facts later, when we prove (p. 51) that the set $S(A)$ of all permutations of A is a *group*.

Example 54. If $A = Z(n) = \{1, \ldots, n\}$, then a permutation of A can be written

$$\theta = \begin{pmatrix} 1 & 2 & & n \\ a_1 & a_2 & \cdots & a_n \end{pmatrix},$$

using the notation of section 3.2. Since θ is assumed to be a bijective map $A \to A$, the elements $a_1, \ldots, a_n$ must be distinct (because θ is injective), and must include all the elements of $Z(n)$ (because θ is surjective). So $a_1, a_2, \ldots, a_n$ must be what is called, in elementary algebra, a 'permutation' of $1, 2, \ldots, n$. Notice that we use this term permutation to refer to the *map* $\theta: A \to A$, and we shall avoid using it in the sense of a re-arrangement $a_1, a_2, \ldots, a_n$ of $1, 2, \ldots, n$.

Example 55. Write $S(n)$ for the set of all permutations of $\{1, 2, \ldots, n\}$. Since there are $n!$ ways of choosing $a_1, a_2, \ldots, a_n$ in the example above, $S(n)$ is a set of $n!$ elements. We shall give below another notation for permutations of a finite set.

Example 56. Let a be a fixed integer, and let θ_a be the map of Z into itself defined by the rule: if $x \in Z$, let $\theta_a(x) = x + a$. Then θ_a is a permutation of Z. For if x, x' are distinct integers, clearly $x + a$, $x' + a$ are distinct; this shows that θ_a is injective. And if y is any integer, there is always an integer x such that $x + a = y$; this shows that θ_a is surjective. Hence $\theta_a: Z \to Z$ is bijective, i.e. is a permutation of Z. Notice that $(\theta_a)^{-1} = \theta_{-a}$, and $\theta_a \theta_b = \theta_{a+b}$; for any $a, b \in Z$.

Example 57. The map $\theta: R \to R$ defined by $\theta(x) = e^x$ is not a permutation of R, because (Example 48, p. 30) it is not surjective.

Cycle notation for permutations

This useful notation can be applied to any permutation $\theta: A \to A$ of a finite set A. We illustrate this notation with the permutation

(a) $\theta = \begin{pmatrix} 1 & 2 & 3 & 4 & 5 & 6 & 7 & 8 & 9 \\ 5 & 3 & 1 & 8 & 2 & 6 & 9 & 4 & 7 \end{pmatrix}$

of the set $Z(9) = \{1, 2, \ldots, 9\}$. In 'cycle notation' θ is written

(b) $\theta = (1523)(48)(6)(79),$

as follows: first take any element of $Z(9)$, say 1. Then θ maps $1 \to 5$, $5 \to 2$, $2 \to 3$, $3 \to 1$. This succession

(c)

is called the *cycle of θ starting with 1*, and is recorded by the *cycle-symbol* (1523). Next take any element of $Z(9)$ which is not in this cycle, and write down the symbol for the cycle of θ starting with this element. For example, we may take the element 4, and then we get the cycle with symbol (48), because θ maps 4 to 8, and 8 back to 4. Next write down the symbol for the cycle starting with any element of $Z(9)$ which does not appear in the cycle-symbols (1523)(48) so far obtained. If we take the element 6, the cycle-symbol is just (6), because θ maps 6 to itself. Finally the cycle starting with 7 has symbol (79), and when we have written this down, our *cycle-expression* (b) for θ is complete, since all the elements of $Z(9)$ appear in it.

Some comments on this notation should be made at this point.

(i) The expression (b) contains the same information as (a), i.e. it tells how θ maps any given element of $Z(9)$. Notice that θ maps the last element of each cycle-symbol in (b), to the first element in that symbol (as for example θ maps $3 \to 1$, $9 \to 7$, etc.).

(ii) The cycle-expression for a given permutation θ is not unique in general. There are two reasons for this. First, the order in which the cycles appear is irrelevant; for example (48)(79)(1523)(6) describes the same permutation θ as (b). The second reason is that each cycle has in general several different cycle-symbols, all equally valid. In fact if the cycle is a *d-cycle* – meaning that it contain d elements – then it has d different cycle-symbols, obtained by starting at each element of the cycle in turn. So the four-cycle (c) has four cycle-symbols (1523), (5231), (2315), (3152), and we could have used any one of these in our cycle-expression for θ. Thus (84)(79)(2315)(6) describes the same permutation θ as (b).

(iii) It is usual to omit one-cycles from the cycle-expressions of given permutation; for example we would omit (6) from (b) and write simply $\theta = (1523)(48)(79)$. It must then be understood that any element which does not appear in a cycle-expression is mapped to itself by the permutation θ. The identity permutation needs a special notation; for example if $A = Z(9)$ then ι_A has cycle-expression (1)(2)(3)(4)(5)(6)(7)(8)(9), and if we omit all one-cycles the expression disappears altogether! For this reason the identity permutation is often written **1**, or ι.

Example 58. The $6 = 3!$ permutations of $Z(3) = \{1, 2, 3\}$ are

$$\begin{pmatrix} 123 \\ 123 \end{pmatrix}, \begin{pmatrix} 123 \\ 312 \end{pmatrix}, \begin{pmatrix} 123 \\ 231 \end{pmatrix}, \begin{pmatrix} 123 \\ 132 \end{pmatrix}, \begin{pmatrix} 123 \\ 321 \end{pmatrix}, \begin{pmatrix} 123 \\ 213 \end{pmatrix}.$$

In cycle notation these are written, respectively, as

1, (132), (123), (23), (13), (12).

Example 59. It is easy to calculate products of permutations in cycle notation. For example, let θ, ϕ be the permutations on the finite set $A = \{a, b, c, d, e\}$, given in cycle notation as

$$\theta = (ac)(be), \qquad \phi = (abc).$$

Then we get a cycle-expression for $\theta\phi$ as follows (remember $\theta\phi$ is the result of applying ϕ, and then θ). $\theta\phi$ maps $a \to e$ (since ϕ maps $a \to b$, and θ maps $b \to e$); then $\theta\phi$ maps $e \to b$ (since ϕ maps $e \to e$, and θ maps $e \to b$); then $\theta\phi$ maps $b \to a$ (since ϕ maps $b \to c$, and ϕ maps $c \to a$). This gives a cycle (aeb). Now take any element of A which has not yet appeared, and calculate the cycle of $\theta\phi$ starting with this element. If we take d as our element, then $\theta\phi$ maps $d \to d$ (since both ϕ and θ map $d \to d$), so the cycle containing d has symbol (d). Finally $\theta\phi$ maps $c \to c$ (since ϕ maps $c \to a$, and θ maps $a \to c$), giving another one-cycle (c). Thus $\theta\phi = (aeb)(d)(c) = (aeb)$.

All products of permutations on $Z(3) = \{1, 2, 3\}$ are given in Table 1 (p. 241).

3.8 Similar sets

Maps play the important role, in all parts of mathematics, of *comparing* different sets. The simplest kind of comparison is that given by the next definition.

DEFINITION. Sets A, B are said to be *similar* if there exists a bijective map $\theta: A \to B$.

We write $A \simeq B$ to denote that A, B are similar. This is a relation between sets, and it has the formal properties of an equivalence relation. If A is any set, then $A \simeq A$, because there is always the identity map $\iota_A: A \to A$, which is bijective. Then if $A \simeq B$ it follows that $B \simeq A$, because if there is a bijective map $\theta: A \to B$, then θ^{-1} is a bijective map $B \to A$. Finally if $A \simeq B$ and $B \simeq C$, then $A \simeq C$. For if $\theta: A \to B$ and $\phi: B \to C$ are given bijective maps, then $\phi\theta: A \to C$ is bijective (see section 3.6(ii), p. 33).

Counting

To say that a finite set A has order n, means that we can *count* the elements $a_1, a_2, \ldots, a_n$ of A and get the 'answer' n. Now this process of counting just consists in setting up a bijective map

$$\theta: \{1, 2, \ldots, n\} \to A,$$

in which $\theta(1) = a_1$, $\theta(2) = a_2$, $\ldots$, $\theta(n) = a_n$. Conversely if A is any set and if θ is any bijection of $\{1, 2, \ldots, n\}$ into A, then A must have order n. For the elements $\theta(1), \ldots, \theta(n)$ are all distinct, since θ is injective; moreover they are *all* the elements of A, since θ is surjective. In other words we can use θ to 'count' the elements of A as $\theta(1), \theta(2), \ldots, \theta(n)$.

Write $Z(n)$ for the set $\{1, 2, \ldots, n\}$. What we have just said shows that A has order n if and only if A is similar to $Z(n)$. Thus similarity has a very simple meaning for finite sets, namely *finite set A, B are similar if and only if they have the same order.*

Similarity is the basis of a theory of 'infinite numbers', or, as they are usually called, 'infinite cardinals'. In this theory, two infinite sets A, B are considered to have the same 'number' of elements, if and only if they are similar. A surprising number of the infinite sets which occur 'naturally' are similar to one of the sets Z or R; but Z is not similar to R, which means that these are infinite sets with different 'numbers' of elements.†

---

†For a proof that Z is not similar to R (this is a non-trivial fact), and a discussion of infinite cardinals, see S. Swierczkowski, *Sets and Numbers*, Routledge & Kegan Paul, London.

Example 60. In Example 48, p. 30, we found a bijective map $\phi: R \rightarrow R^+$, which shows that R is similar to R^+. So it is possible for an infinite set to be similar to a proper subset of itself. For a *finite* set A this is impossible, because any proper subset B of A has smaller order than the order of A.

Example 61. Let $B = \mathscr{B}(X)$ be the set of all subsets of a set X of order n. We shall prove that B has order 2^n (we have already proved this by a different method, Example 35, p. 19) by proving that B is similar to the set M of all maps of X into the two-element set $U = \{0, 1\}$. This will prove that B has the same order as M, and we know from section 3.2 (p. 28) that M has 2^n elements.

To prove that B is similar to M, we must define a bijective map $\theta: B \rightarrow M$. We do this as follows. For any $S \in B$, i.e. for any subset S of X, we define a map $\chi_S: X \rightarrow U$ by the rule: for any $x \in X$, define $\chi_S(x) = 1$ if $x \in S$, and $\chi_S(x) = 0$ if $x \notin S$. Now we define $\theta: B \rightarrow M$ by the rule: if $S \in B$, let $\theta(S) = \chi_S$. It is easy to verify that θ is bijective, and we leave this as an exercise for the reader.

Exercises for Chapter 3

1 Write down all the maps of the set $B = \{a, b\}$ into $A = \{1, 2, 3\}$. How many of these are injective? Find a formula for the number of injective maps of a set of order r into a set of order $s (r \leq s)$.

2 Which of the following maps $\theta: Z \rightarrow Z$ are (a) injective, (b) surjective? In each case $\theta(x)$ is given for any $x \in Z$. (i) $\theta(x) = x^2$; (ii) $\theta(x) = -x$; (iii) $\theta(x) = \frac{1}{2}x$ if x is even, $\theta(x) = 0$ if x is odd; (iv) $\theta(x) = 2x + 1$.

3 Let S be the set of all real numbers $x \geq 0$, and define $\theta: S \rightarrow S$ by $\theta(x) = x^2$. Prove that θ is bijective, and find its inverse. Now define $\phi: R \rightarrow R$ by $\phi(x) = x^2$. Is ϕ bijective?

4 Let a, b be fixed real numbers, and define $\theta: R \rightarrow R$ by the formula $\theta(x) = ax + b$. Prove that θ is bijective if and only if $a \neq 0$, and give a formula for θ^{-1} in this case.

5 Write $\theta_{a,b}$ for the map of the preceding exercise. Prove that $\theta_{a,b}\theta_{c,d} = \theta_{ac, ad+b}$, for any real numbers a, b, c, d. For which values of a, b, c, d do $\theta_{a,b}$ and $\theta_{c,d}$ commute?

6 If $\theta: A \rightarrow B$ and $\phi: B \rightarrow A$ satisfy $\theta\phi = \iota_B$, does it follow that θ is bijective? (Try some examples of maps θ, ϕ with A, B as in

Exercise 1.) Prove that if $\theta\phi = \iota_B$, then the map $\alpha = \phi\theta$ satisfies the equation $\alpha^2 = \alpha$.

7 Express the permutations α, ρ of Example 50, p. 31 in cycle notation. Calculate the products $\alpha\rho$, $\rho\alpha$ in cycle notation using the technique of Example 59, and check your answer against the expressions for these maps given in Example 50.

8 Express the permutation

$$\theta = \begin{pmatrix} a & b & c & d & e & f & g \\ a & f & e & g & c & d & b \end{pmatrix}$$

in cycle notation.

9 Let α be the permutation of $Z(n) = \{1, 2, \ldots, n\}$ given in cycle notation as $\alpha = (a_1 a_2 \ldots a_p)$. Prove that $\alpha^p = 1$. (Notice that $a_1, a_2, \ldots, a_p$ must be distinct elements of $Z(n)$. Also, by the convention (iii) of p. 00, we have $\alpha(j) = j$, for any $j \in Z(n)$ which is not in the set $\{a_1, \ldots, a_p\}$. It is a good idea to try some special cases, e.g. take $n = 5$ and $\alpha = (1342)$, and calculate, in turn, the maps α^2, α^3, α^4.)

10 Let α, β be permutations of $Z(n)$ given in cycle notation as $\alpha = (a_1 a_2 \ldots a_p)$, $\beta = (b_1 b_2 \ldots b_q)$. Prove that if these cycles are *disjoint*, meaning that

$$\{a_1, a_2, \ldots, a_p\} \cap \{b_1, b_2, \ldots, b_q\} = \varnothing,$$

then α, β commute. Prove also that $\alpha\beta$ has cycle expression $(a_1 a_2 \ldots a_p)(b_1 b_2 \ldots b_q)$. (This result can be generalized to a product of k disjoint cycles ($k \geq 3$). Hence the 'cycle-expression' of a permutation can always be regarded as a product of disjoint cycles – for example the permutation θ on p. 36 is the product of the cycles (1523), (48), (79).)

11 Show that if A is a finite set, then any map $\theta: A \rightarrow A$ which is injective, is also surjective. Find an example in this chapter which shows this to be false for an infinite set A. Construct an example of a map $\theta: A \rightarrow A$ which is surjective but not injective. Can this be done with A finite?

12 Show that the set of all even integers is similar to the set of all integers.

13 If A, B, C are any sets, prove (i) $A \times B \simeq B \times A$, and (ii) $(A \times B) \times C \simeq A \times (B \times C)$.

14 If X, Y are two disjoint sets, prove that $\mathscr{B}(X \cup Y) \simeq \mathscr{B}(X) \times \mathscr{B}(Y)$. (Define $\theta : \mathscr{B}(X \cup Y) \to \mathscr{B}(X) \times \mathscr{B}(Y)$ by the rule: if $S \in \mathscr{B}(X \cup Y)$, let $\theta(S) = (S \cap X, S \cap Y)$. Prove that θ is bijective.)

15 If $A \simeq B$, prove that $S(A) \simeq S(B)$ (here $S(C)$ denotes the set of all permutations of C, for any given set C).

4 Groups

4.1 Binary operations on a set

If x and y are ordinary numbers, there are various ways of combining or 'operating with' x and y to give another number, for example we may form their sum $x+y$, or difference $x-y$, or product xy. These are three examples of *binary operations* on the set R of all real numbers.

Modern algebra is largely concerned with the abstract properties of operations like this. In general, a binary operation ω on an arbitrary set A is nothing more than a map of the set $A \times A$ into A, so that ω maps each *pair* (x, y) of elements x, y of A, to some element $\omega(x, y)$ of A.† For example, the operation of addition is the map $\alpha: R \times R \to R$ which maps (x, y) to the element $\alpha(x, y) = x+y$ of R. Similarly, subtraction is the map $\beta: R \times R \to R$ defined by the rule: $\beta(x, y) = x-y$. Multiplication is the map $\gamma: R \times R \to R$ defined by the rule: $\gamma(x, y) = xy$.

A general binary operation $\omega: A \times A \to A$ can be thought of as a kind of generalized 'multiplication' or 'addition' on A. This point of view is emphasized if we use a notation such as $x \circ y$ in place of $\omega(x, y)$; if we do this, we can in fact dispense with the symbol ω altogether, and refer simply to 'the binary operation $\circ$'. With this notation our definition runs as follows.

DEFINITION. Let A be any set. Then a *binary operation $\circ$ on A* is a map of $A \times A$ into A, which maps each pair (x, y) of elements of A to an element $x \circ y$ of A.

---

†With our standard notation for maps, this should be $\omega((x, y))$. But it is usual to omit one pair of parentheses, for clarity.

Additive and multiplicative notations

It is very common, particularly in group theory, to use one of the familiar notations $x + y$ or xy, in place of $x \circ y$, even when x, y are not numbers and $\circ$ is not one of the ordinary arithmetical operations. We say then that we are using the *additive*, or *multiplicative*, *notation* for the operation in question. We would then usually refer to the operation itself as 'the operation $+$' (if additive notation is being used) or 'the operation $.$' (if multiplicative notation is being used – we are in some embarrassment here, since the notation xy uses no symbol between x and y! In this book we shall keep to the convention that if a binary operation is called $.$, then it maps the pair (x, y) to xy, and *not* to $x . y$).

Example 62. If A is a small finite set, then a binary operation on A can be represented by its *multiplication table*. The table below represents a binary operation $\circ$ on $A = \{a, b, c\}$, for which $a \circ a = b$, $a \circ b = c$, $a \circ c = b$, $b \circ a = b$, etc.

	a	b	c
a	b	c	b
b	b	a	c
c	a	c	c

If we decided to use the additive notation for this operation, we should write $a + a = b$, $a + b = c$, etc. In multiplicative notation, these would be written $aa = b$, $ab = c$, etc.

Example 63. Let $S(A)$ be the set of all permutations of a given set A. Then the product of maps defines a binary operation on $S(A)$, i.e. if θ, $\phi \in S(A)$, we take $\theta \circ \phi$ to be $\theta\phi$, which again belongs to $S(A)$. Notice that we are using the convention of multiplicative notation here.

Example 64. Let $\mathcal{B}(X)$ be the set of all subsets of a given set X (Example 25, p. 12). If A, $B \in \mathcal{B}(X)$, i.e. if A, B are subsets of X, then $A \cap B$ and $A \cup B$ also belong to $\mathcal{B}(X)$. In this way we have two binary operations $\cap$ and $\cup$ on $\mathcal{B}(X)$. The *difference* $A - B$ (p. 8) gives another binary operation on $\mathcal{B}(X)$.

4.2   Commutative and associative operations

In this section and the next we describe some special kinds of operation, which are relevant to the definition of a group (see section 4.4).

Commutative operations

DEFINITION. If ∘ is a binary operation on a set A, and if x, y are elements of A, we say that x and y commute (with respect to ∘) if $x \circ y = y \circ x$. The operation ∘ is called *commutative* if it satisfies the *commutative law*

$$x \circ y = y \circ x \tag{4.1}$$

for all x, $y \in A$, i.e. if every pair of elements of A commute.

Example 65. In additive or multiplicative notations, equation (4.1) reads

$$x + y = y + x,$$

or

$$xy = yx,$$

respectively. Since these are both true when x, y are ordinary numbers, it means that the operations of ordinary addition and multiplication are both commutative. The operation of subtraction is not commutative, because if we take $x \circ y = x - y$, then equation (4.1) reads $x - y = y - x$, which is not true for all x, y.

Example 66. We saw in Example 50, p. 31 an example of two mappings α, ρ, which are in fact permutations of the set $\{1, 2, 3\}$, which do not commute. This shows that the binary operation of Example 63 is not, in general, commutative.

Example 67. The operations $\cap$ and $\cup$ of Example 64 are both commutative, since $A \cap B = B \cap A$ and $A \cup B = B \cup A$, for any A, $B \in \mathscr{B}(X)$.

Associative operations

DEFINITION. A binary operation ∘ on a set A is called *associative* if it satisfies the *associative law*.

$$(x \circ y) \circ z = x \circ (y \circ z) \qquad (4.2)$$

for all $x, y, z \in A$.

Example 68. In additive or multiplicative notations equation (4.2) reads

$$(x + y) + z = x + (y + z),$$

or

$$(xy)z = x(yz),$$

respectively. Both of these are true when x, y, z are ordinary numbers, which shows that the operations of ordinary addition and multiplication are both associative. Subtraction of numbers is not associative, because if we take $x \circ y = x - y$, equation (4.2) reads $(x - y) - z = x - (y - z)$, which is not true for all x, y, z.

Example 69. The operation of mapping product on $S(A)$ (Example 63) is associative, by the associative law for mappings (p. 32). Thus it is possible to have an associative operation, which is not commutative (Example 66).

Example 70. The operations $\cap$ and $\cup$ of Example 64 are both associative (Example 27, p. 13).

Example 71. The operation of Example 62, p. 43, is neither commutative nor associative. For $a \circ b \neq b \circ a$, and therefore it is not commutative. It is not associative; for example $(a \circ a) \circ b \neq a \circ (a \circ b)$ (the left side is $b \circ b = a$, and the right side is $a \circ c = b$).

General associative law

Suppose for the moment that $\circ$ is any binary operation on a set A. If we want to work out a product

$$x_1 \circ x_2 \circ \ldots \circ x_n \qquad (4.3)$$

of n factors $x_1, x_2, \ldots, x_n \in A$, we must do this as a succession of products of *pairs* of elements of A, and we can show by putting in brackets, how this is to be done. For example there are five ways of working out a product of four factors x, y, z, w, *viz.* $x \circ [y \circ (z \circ w)]$, $x \circ [(y \circ z) \circ w]$, $(x \circ y) \circ (z \circ w)$, $[(x \circ y) \circ z] \circ w$, $[x \circ (y \circ z)] \circ w$. In general, these would all be different. But if $\circ$ is *associative*, then they will all be the same.

THEOREM (*General associative law*). *If $\circ$ is an associative binary operation on A, and if $x_1, x_2, \ldots, x_n$ are given elements of A, then the product* (4.3) *has the same value, however it is 'bracketed'.*

This means that when $\circ$ is associative, there is no need to put brackets into a product like (4.3).

Proof of the general associative law. We prove this by induction on n. For $n = 1, 2$ there is no problem, because there is only one way of working out (4.3) in these cases. For $n = 3$ there are two bracketings, *viz.* $(x_1 \circ x_2) \circ x_3$ and $x_1 \circ (x_2 \circ x_3)$, and of course these *are* equal, because we are assuming that $\circ$ satisfies the associative law. Now assume $n > 3$, and as induction hypothesis, that products of fewer than n factors are independent of bracketing. In whatever way product (4.3) is worked out, the last step will be to make a product of the form

$$P_r = (x_1 \circ \cdots \circ x_r) \circ (x_{r+1} \circ \cdots \circ x_n),$$

for some value of r in the range $1, 2, \ldots, n-1$. In P_r it is assumed that $x_1 \circ \cdots \circ x_r$ and $x_{r+1} \circ \cdots \circ x_n$ have already been worked out – by the induction hypothesis, these products are independent of the way they were bracketed. Now we complete our proof by showing that $P_1 = P_2 = \cdots = P_{n-1}$, for we know that product (4.3), however it is bracketed, is equal to one of the P_r. For any r in the range $1, 2, \ldots, n-2$ put $x = x_1 \circ \cdots \circ x_r$, $y = x_{r+1}$ and $z = x_{r+2} \circ \cdots \circ x_n$. Then equation (4.2) gives $(x \circ y) \circ z = x \circ (y \circ z)$, i.e. $P_{r+1} = P_r$. Hence $P_{n-1} = P_{n-2} = \cdots = P_2 = P_1$, as required.

Powers

Let $\circ$ be an associative binary operation on a set A. If we take $x_1 = x_2 = \cdots = x_n$ in product (4.3), and if we write $x = x_1$, then product (4.3) is called the *nth power* $x \circ x \circ \cdots \circ x$ of x. In multiplicative notation (and also often in the general notation) this is written x^n. In additive notation the nth power is written nx, because if we take A to be the set of all real numbers and $+$ as ordinary addition, then $x + x + \cdots + x$, with n terms x, is in fact the ordinary product nx. In the case of a *general* associative operation

+ we still use the notation nx, but of course this should no longer be thought of as a product.

Just as for powers of a map (p. 32) we have, for any positive integers m, n, the 'index law'

$$x^m \circ x^n = x^{m+n}, \tag{4.4}$$

because $x^m \circ x^n$ is just a product of $m+n$ factors x. It follows that $x^m \circ x^n = x^n \circ x^m$, i.e. *any two powers of a given element commute*, although we are not assuming that $\circ$ is a commutative operation.

Example 72. Powers, beyond the second, cannot even be defined for a general binary operation. Let $\circ$ be the operation of Example 62, p. 43. If we try to define a^3, we have two choices, *viz.* $(a \circ a) \circ a$ and $a \circ (a \circ a)$, and these are not equal (they are $b \circ a = b$ and $a \circ b = c$, respectively). This indicates the kind of difficulty encountered in studying non-associative operations.

Example 73. In additive notation, the index law (4.4) becomes

$$mx + nx = (m+n)x.$$

Example 74. If the operation $\circ$ is *both* associative and commutative, then the factors in the product (4.3) can be permuted in any way without altering its value. In particular it is easy to prove, for any x, $y \in A$ and any positive integer n, that $(x \circ y)^n = x^n \circ y^n$, for each side is simply a product of n factors equal to x with n factors equal to y. But if $\circ$ is not commutative, this fails (see Exercise 9, p. 55).

4.3 Units and zeros

Let A be any set, and $\circ$ a binary operation on A.

DEFINITION 1. Any element e of A which satisfies

$$e \circ x = x \circ e = x, \tag{4.5}$$

for all x in A, is called a *unit element* (or a *neutral element*) for the operation $\circ$.

There is no need for such a unit element to exist, as we shall see (Example 76, below). But if A does contain a unit element e for the operation $\circ$, then it is *unique*, i.e. we have the following theorem.

THEOREM. *If e, f are both unit elements for ∘, then e = f.*

Proof. We put $x = f$ in equation (4.5) and get $e \circ f = f$. But f satisfies $f \circ x = x \circ f = x$ for all x in A, so putting $x = e$, we get in particular $e \circ f = e$. Thus $e = e \circ f = f$.

Example 75. The number 1 is a unit element for ordinary multiplication, because $1x = x1 = x$ for all x in R. There is also a unit element for the ordinary operation of addition, but this is not 1, but the number 0. For if we take $x \circ y$ to mean $x + y$, then equation (4.5) becomes $e + x = x + e = x$, which is satisfied, for all x in R, by $e = 0$.

Example 76. There is no unit for the ordinary operation of subtraction. For this would have to be a number e such that $e - x = x - e = x$ for all x in R, which is impossible. The operation $\circ$ of Example 62 has no unit.

Example 77. The identity permutation ι_A is a unit element for the operation of mapping product on $S(A)$ (Example 63, p. 43), for $\iota_A \theta = \theta \iota_A = 0$ for all θ in $S(A)$ (see p. 33).

Zeros

A unit element e 'leaves alone' any element x by which it is 'multiplied'. As an opposite extreme, we define next elements which 'swallow up' every other element.

DEFINITION 2. Any element n of A which satisfies

$$n \circ x = x \circ n = n, \tag{4.6}$$

for all x in A, is called a *zero element* (or *annihilator*) for the operation $\circ$.

Just as for units, there is no need for a given binary operation to have a zero element, but if it does have one, then this is unique, i.e. we have the following theorem.

THEOREM. *If n, p are both zero elements for ∘, then n = p.*

Proof. We put $x = p$ in equation (4.2) and get $n \circ p = n$. But p satisfies $p \circ x = x \circ p = p$ for all x in A, so putting $x = n$, we get in particular $n \circ p = p$. Thus $n = n \circ p = p$.

Example 78. The number 0 is a zero element for ordinary multiplication, because $0x = x0 = 0$ for all x in R. But there is no zero element for the ordinary operation of addition, for this would have to be a number n such that $n + x = x + n = n$ for all x in R, and this is impossible.

Example 79. The subsets X, $\varnothing$ of a given set X are the unit and zero elements, respectively, for the operation $\cap$ on $\mathscr{B}(X)$ (Example 64, p. 43). For $X \cap A = A \cap X = A$, and $\varnothing \cap A = A \cap \varnothing = \varnothing$, for all subsets A of X. For the operation $\cup$, these roles are reversed; X is the zero and $\varnothing$ the unit element.

4.4 Gruppoids, semigroups and groups

In this section we shall define the most elementary *algebraic structures*. An algebraic structure consists of a *set* (usually assumed to be non-empty), together with one or more *operations* on this set. For example, if G is any set and $\circ$ is any binary operation on G, then the pair $(G, \circ)$ is called a *gruppoid*. This is the first type of algebraic structure we shall consider.

DEFINITION 3. A gruppoid is a pair $(G, \circ)$, where G is a non-empty set, and $\circ$ is a binary operation on G.

If G is a finite set, we say that $(G, \circ)$ is a *finite gruppoid*, with *order* $|G|$. Otherwise $(G, \circ)$ is an *infinite gruppoid*.

Example 80. A gruppoid $(G, \circ)$ consists of the two 'components', a *set G*, and a *binary operation* $\circ$ on G. There can be many different gruppoids with the same set. For example if R is the set of all real numbers, then $(R, +)$, $(R, -)$ and $(R, .)$ (here . stands for ordinary multiplication) are three different gruppoids.

If $\circ$ is the operation of Example 62, p. 43, then $(A, \circ)$ is a finite gruppoid of order 3. All the examples of binary operations which we have given, provide examples of gruppoids.

DEFINITION 4. A *semigroup* is a gruppoid $(G, \circ)$ whose operation $\circ$ is associative, i.e. which satisfies the associative law

G1 $(x \circ y) \circ z = x \circ (y \circ z),$ for all x, y, z in G.

Example 81. $(R, +)$ and $(R, .)$ are semigroups, but not $(R, -)$ (see

Example 68, p. 45). $[\mathscr{B}(X), \cap]$ and $[\mathscr{B}(X), \cup]$ are both semigroups (Example 70, p. 45).

DEFINITION 5. A *group* is a semigroup $(G, \circ)$ which satisfies, in addition to **G1**, two further conditions

G2 G has a *unit element e*, i.e.

$e \circ x = x \circ e = x,$ for all x in G, and

G3 every element of G has an *inverse*, i.e. for each x in G there is an element $\hat{x}$ in G, called the inverse of x, such that

$x \circ \hat{x} = \hat{x} \circ x = e.$

Thus *a group is a set G, together with a binary operation ∘ on G, such that* **G1**, **G2** *and* **G3** *hold*. The conditions **G1, G2, G3** are called the *group axioms*; the theorems of *group theory* are those which can be deduced from these axioms.

Every group $(G, \circ)$ is of course also a semigroup, and every semigroup is a gruppoid. So gruppoids are the most 'general' of the three types of structure we have defined. Any theorems about arbitrary gruppoids would have very wide applications, because examples of gruppoids occur in mathematics at every turn. Unfortunately very few non-trivial theorems about gruppoids have been found, so that, although very general, gruppoid theory is also very dull. The introduction of the associative law **G1** makes more progress possible, but it is the combination of all three assumptions **G1, G2, G3** which gives group theory a unique position; it is general enough to draw ideas and interest from all parts of mathematics, but at the same time special and detailed enough to have its own 'personality'. For the rest of this chapter, and in the three following chapters, we shall discuss some of these ideas or concepts of group theory.

Abelian groups

The commutative law is not one of the axioms of group theory. But if it happens that a group $(G, \circ)$ does satisfy this law, i.e. if

G4 $x \circ y = y \circ x,$ for all x, y in G,

then $(G, \circ)$ is called an *Abelian* (or *commutative*) group. (This adjective Abelian is taken from the name of the Norwegian mathematician N. H. Abel (1802–29).) Thus an Abelian group is a gruppoid which satisfies **G1**, **G2**, **G3** and **G4**.

4.5   Examples of groups

Example 82. One-element groups. If a group $(G, \circ)$ has only one element, then, since it must have a unit e, $G = \{e\}$, and the operation $\circ$ is completely defined by $e \circ e = e$. Conversely if we take any one-element set $G = \{e\}$ and define $\circ$ in this way, then it is easy to check **G1**, **G2**, **G3** and verify that $(G, \circ)$ is a group.

Example 83. Symmetric groups. Let A be any given set, and $S(A)$ the set of all permutations of A. If we denote the binary operation giving the mapping product $\theta\phi$ by . , then $[S(A), .\,]$ is a group, called the *symmetric group* on A. We verify this by checking the group axioms.

   G1 holds, since mapping product is associative (p. 32).

   G2 holds, taking $e = \iota_A$ (Example 77, p. 48).

   G3 holds, because if $\theta \in S(A)$, then $\theta^{-1} \in S(A)$ and satisfies $\theta\theta^{-1} = \theta^{-1}\theta = \iota_A$ (p. 33), i.e. we can take $\hat{\theta} = \theta^{-1}$.

   When A is the set $\{1, 2, \ldots, n\}$, we write $S(A) = S(n)$. The group $[S(n), .\,]$ is finite, of order $n!$ (Example 55, p. 35); it is called the *symmetric group of degree n*. The 'multiplication table' for $[S(3), .\,]$ is given on p. 241. As we have already seen (Example 66, p. 44), this is not an Abelian group if $n \geq 3$.

Example 84. The additive group of integers. Let Z be the set of all integers, and let $+$ denote ordinary addition. Then $(Z, +)$ is a group. **G1** holds, i.e. $(x + y) + z = x + (y + z)$ for all integers x, y, z. **G2** holds if we take $e = 0$. **G3** holds if we take $\hat{x} = -x$. This is an Abelian group, because it satisfies also the commutative law $x + y = y + x$ for all $x, y \in Z$.

Example 85. The additive group of reals. Let R be the set of all real numbers and $+$ the ordinary operation of addition. Then $(R, +)$ is an Abelian group, just as in the last example. The unit element is again 0, and the 'inverse' $\hat{x}$ of x is $-x$.

Example 86. The multiplicative group of reals. $(R, .\,)$ is a semigroup, but not a group. For **G1** holds, $(xy)z = x(yz)$ for all x, y, $z \in R$, and in fact **G2** holds, taking $e = 1$. But there is no number $\hat{0}$ satisfying $0\hat{0} = \hat{0}0 = 1$, so that

0 does not have an inverse, and **G3** fails. However, if we take instead of R the set R^* of all *non-zero* real numbers, then $(R^*, .)$ is a group (notice that if $x \in R^*$, $y \in R^*$ then $xy \in R^*$, i.e. multiplication does define a binary operation on R^*). In fact **G1, G2** hold just as before, and also **G3**, since for any $x \in R^*$ we take $\hat{x} = 1/x$. $(R^*, .)$ is an Abelian group.

Example 87. A group $(G, \circ)$ which has a zero element n must be a one-element group. For if x is any element of G, $x = x \circ e = x \circ (n \circ \hat{n}) = (x \circ n) \circ \hat{n} = n \circ \hat{n} = e$. Thus $G = \{e\}$.

Additive and multiplicative notations for groups

It is usual in group theory to use either the additive or multiplicative notations for the group operation $\circ$. There are some standard conventions, which we now describe.

Additive notation
In this notation we write

$x + y$,	for $x \circ y$ (and call this the *sum* of x and y),
0	for the unit element e (but notice, this is *not* a zero element),
$-x$	for the inverse $\hat{x}$ of an element x of G,
$x - y$	for $x \circ \hat{y}$,
nx	for the nth power $x \circ x \circ \cdots \circ x$, n a positive integer.

The notation nx is extended to arbitrary integers by writing $0x = 0$ and $(-n)x = n(-x) = -(nx)$ (see section 4.6(v)).

Multiplicative notation
In this notation we write

xy	for $x \circ y$ (and call this the *product* of x and y),
e or 1	for the unit element,
x^{-1}	for the inverse $\hat{x}$,
x^n	for the nth power $x \circ x \circ \cdots \circ x$, n a positive integer.

The notation x^n is extended to arbitrary integers by writing $x^0 = e$ and $x^{-n} = (x^{-1})^n = (x^n)^{-1}$ (see section 4.6(v)).

The additive notation is almost always reserved for Abelian groups (e.g. $(Z, +)$, Example 84).

4.6 Elementary theorems on groups

We collect in this section the simplest theorems about groups. Like all the theorems of group theory, they are proved using only the axioms **G1**, **G2** and **G3**. It does not matter which convention of notation we use, and we shall use multiplicative notation. Everything could be 'translated' into additive notation if we preferred (see Example 88, below).

Let $(G, .)$ be a group, e its unit element. By the theorem on p. 48, *e is unique*. Let us write x^{-1} for the inverse x of a given element x of G; axiom **G3** says that such an element exists and satisfies (a) $x^{-1}x = e$, and (b) $xx^{-1} = e$. The next theorem shows that *the inverse of x is unique*, in fact that x^{-1} is the only element of G which satisfies either of the conditions (a) and (b).

(i) THEOREM. *If $x \in G$ and if a is any element of G such that $ax = e$, then $a = x^{-1}$. Similarly if b is any element such that $xb = e$, then $b = x^{-1}$.*

We shall prove this as a consequence of the following more general theorem.

(ii) THEOREM. *If x, $y \in G$ then there is one and only one element a of G such that $ax = y$, namely $a = yx^{-1}$. Similarly there is one and only one element b of G such that $xb = y$, namely $b = x^{-1}y$.*

Proof. If $ax = y$, multiply on the right by x^{-1}, giving $(ax)x^{-1} = yx^{-1}$. But $(ax)x^{-1} = a(xx^{-1})$ (by **G1**) $= ae$ (by **G3**) $= a$ (by **G2**); therefore $ax = y$ implies $a = yx^{-1}$. Conversely if $a = yx^{-1}$, then $ax = (yx^{-1})x = y(x^{-1}x) = ye = y$. We leave to the reader the corresponding task of proving that $xb = y$ if and only if $b = x^{-1}y$. To prove theorem (i), we now have only to put $y = e$ in theorem (ii).

(iii) THEOREM. *If $x \in G$ then $(x^{-1})^{-1} = x$.*

Proof. Since $xx^{-1} = e$ by **G3**, we get $x = (x^{-1})^{-1}$ from (i) (replace a in (i) by x, and x by x^{-1}).

(iv) THEOREM. *If x, $y \in G$ then $(xy)^{-1} = y^{-1}x^{-1}$.*

Proof. Replace a in (i) by $y^{-1}x^{-1}$ and x by xy. Since $(y^{-1}x^{-1})(xy) = [y^{-1}(x^{-1}x)]y$ by the general associative law (p. 45), this product is equal to $(y^{-1}e)y = y^{-1}y = e$. Hence (i) gives $y^{-1}x^{-1} = (xy)^{-1}$.

This result extends easily to a product of n factors, $(x_1 x_2 \ldots x_n)^{-1} = x_n^{-1} \ldots x_2^{-1} x_1^{-1}$. Taking $x_1, x_2, \ldots, x_n$ all equal to x we get a proof of the following theorem.

(v) THEOREM. *If $x \in G$ and n is any positive integer, $(x^n)^{-1} = (x^{-1})^n$.* This justifies the notation, which we have already introduced, by which we write x^n for $(x^n)^{-1}$; it is the same as $(x^{-1})^n$.

Example 88. In additive notation, theorem (ii) would read: 'If x, $y \in G$ there is one and only one element a of G such that $a + x = y$, namely $a = y + (-x)$ (or $y - x$). Similarly, there is one and only one element b of G such that $x + b = y$, namely $b = (-x) + y$.

Example 89. The inverse of e is e. (Take $a = x = e$ in theorem (i); $ee = e$ by axiom **G2**.)

Example 90. Each element x commutes with its inverse, because axiom **G3** says $xx^{-1} = x^{-1}x = e$.

Example 91. General index law. If m, n are positive integers

(vi) $x^m x^n = x^{m+n}$, for any $x \in G$,

by the associative law (p. 47); this still holds if either m, n is zero, by our convention $x^0 = e$. We show now that (vi) holds for arbitrary integers m, n. As we have seen, it does hold if (a) $m \geq 0$ and $n \geq 0$. Next we prove it in the case (b) $m \geq 0$ and $m + n \geq 0$. We have only to consider the case $n < 0$, say $n = -s(s > 0)$. Then by (a) $x^{m+n}x^s = x^{m+n+s} = x^m$. Now multiply both sides on the right by $(x^s)^{-1} = x^n$ and we get (vi). A similar proof holds in case (c) $n \geq 0$ *and* $m + n \geq 0$. Suppose finally (d) *No two of m, n, $m + n$ are ≥ 0.* Then certainly at least two of $-n$, $-m$, $-n - m$ are positive, and we can apply one of the previous cases to prove $x^{-n}x^{-m} = x^{-n-m}$. Now take inverses of both sides, and we get (vi). So (vi) holds for any integers m, n. By a similar discussion we may prove also (vii) $(x^m)^n = x^{mn}$ *for any $x \in G$, and any integers m, n.*

Exercises for Chapter 4

1 Show that there are n^{n^2} different binary operations on a set A of order n. Find all sixteen binary operations on $A = \{a, b\}$. How many are (i) commutative, (ii) associative, (iii) have unit elements, or (iv) have zero elements?

2 Define $\circ$ on R by the rule: if $x, y \in R$ then $x \circ y = xy + 1$. Show that this is commutative but not associative. Does it have unit or zero elements?

3 If $\circ$ is a binary operation on a set A, and if an element z is both a unit and a zero for $\circ$, show that $A = \{z\}$.

4 An element e which satisfies $e \circ x = x$ for all x in A is a 'left unit' for $\circ$. Show that left units are not necessarily unique and (hence) they may not be unit elements, by considering the following example: let $A = \{a, b\}$ and define $\circ$ by $a \circ a = b \circ a = a$, $a \circ b = b \circ b = b$.

5 Define right units similarly, and show that if $\circ$ has a right unit e and a left unit f, then $e = f$ and this is a unit element.

6 Make similar definitions and prove results analogous to the above for left and right zeros of a binary operation.

7 Let $M(A)$ denote the set of all maps of a set A into itself. Show that $[M(A), .]$ is a semigroup, where . denotes product of maps. Is this a group?

8 Let $(G, .)$ be a semigroup and x, y elements of G which commute. Show that x^m, y^n commute, where m, n are any positive integers.

9 Taking α, ρ as in Example 50, p. 31, show that $(\alpha\rho)^2 \neq \alpha^2\rho^2$.

10 Define maps α, β of Z into itself as follows: for any $x \in Z$, $\alpha(x) = 2x$; while $\beta(x) = \frac{1}{2}x$ if x is even, and $\beta(x) = 0$ if x is odd. Show that $\beta\alpha = \iota$ (the identity map on Z), but $\alpha\beta \neq \iota$. Prove that if $(G, .)$ is any semigroup with unit element e, and if $a, b \in G$ are such that $ba = e$, then $(ab)^2 = ab$. Verify this in the example just given.

11 Define a binary operation $\circ$ on R by $x \circ y = x + y + xy$. Prove that $(R, \circ)$ is a semigroup, with unit element 0. If R' is the set of real numbers $x \neq -1$, show that $(R', \circ)$ is a group.

12 Show that $[\mathscr{B}(X), \oplus]$ is an Abelian group, where $\mathscr{B}(X)$ is the set of all subsets of a set X, and $\oplus$ is defined in Exercise 7, p. 14.

13 Let $(G, .)$ be a semigroup with unit element e, and say that $x \in G$ is *invertible* if there exists some $y \in G$ such that $xy = yx = e$. Prove that if x, x' are invertible, so is xx', and show that $(G^*, .)$ is a group, where G^* is the set of all invertible elements. Find the order of this group when $(G, .) = (Z, .)$.

14 If x, y are elements of a group and $x^2 = y^2 = (xy)^2 = e$, prove that x and y commute.

15 Let G be the set of all maps $\theta_{a,b}$ of Exercises 4 and 5, p. 39, such that $a \neq 0$. Prove that $(G, .)$ is a non-Abelian group, where . denotes product of maps.

5 Subgroups

5.1 Subsets closed to an operation

Let G be a set, and $\circ$ a binary operation on G.

DEFINITION. A non-empty subset H of G is said to be *closed to* $\circ$ if, for every pair (x, y) of elements of H, also $x \circ y$ belongs to H; i.e. if H satisfies the condition

S1 $x, y \in H \Rightarrow x \circ y \in H$.

Example 92. Let Z be the set of all integers, and Y the subset consisting of all integers $x \geq 0$. Then Y is closed to both the operations of addition and multiplication, because if $x, y \in Y$ then both $x + y$ and xy belong to Y. But Y is not closed to subtraction (for example 0, $1 \in Y$ but $0 - 1 \notin Y$).

If H is closed to $\circ$, we can use $\circ$ as a binary operation on H,† so that $(H, \circ)$ is a gruppoid which is a part of the gruppoid $(G, \circ)$; we say sometimes that H is a *subgruppoid* of $(G, \circ)$. If $\circ$ is associative as operation on G, then it is associative as operation on H (for if $(x \circ y) \circ z = x \circ (y \circ z)$ for all x, y, z in G, then certainly this holds for all x, y, z in H!); similarly if $\circ$ is commutative on G. In particular if H is a subgruppoid of a *semigroup* $(G, \circ)$, then $(H, \circ)$ is itself a semigroup. But Example 92 shows that if $(G, \circ)$ is a *group* and H is closed to $\circ$, then $(H, \circ)$ need not be a group − for $(Z, +)$ is a group, Y is closed to $+$, but $(Y, +)$ is not a group since it does not contain the inverse $-y$ of each element $y \in Y$.

---

†It would be more accurate, but perhaps tedious, to use separate symbols for $\circ$ as operation on G, which is a map $G \times G \to G$, and for $\circ$ as operation on H, which is a map $H \times H \to H$.

5.2   Subgroups

DEFINITION. Let $(G, \circ)$ be a group, and H a non-empty subset of G. Then H is called a *subgroup of* $(G, \circ)$ if

S1 $x, y \in H \Rightarrow x \circ y \in H,$

and

S2 $x \in H \Rightarrow \hat{x} \in H.$

Thus H has to be closed to $\circ$, and also to the 'operation' of taking inverses.

THEOREM. *If H is a subgroup of $(G, \circ)$, then $(H, \circ)$ is a group.*

Proof. $(H, \circ)$ satisfies the associative law **G1**, as we have seen. Then it also satisfies **G2**, because in fact we have the following lemma.

LEMMA. *Any subgroup H of $(G, \circ)$ contains the unit element e of G.*

To prove this lemma, take any $x \in H$. By **S2**, $\hat{x} \in H$. Then using **S1**, $x \circ \hat{x} = e \in H$.

Finally $(H, \circ)$ satisfies **G3**, by condition **S2**. This proves the theorem.

The two conditions **S1**, **S2** can be put as a single condition, as follows: *a non-empty subset H of G is a subgroup if and only if*

S $x, y \in H = x \circ \hat{y} \in H.$

First suppose H is a subgroup, i.e. satisfies **S1** and **S2**. Let $x, y \in H$. By **S2**, $\hat{y} \in H$; now by **S1**, $x \circ \hat{y} \in H$, this verifies **S**. Conversely suppose H satisfies **S**. Take any $x \in H$, and put $y = x$ in **S**, which gives $x \circ \hat{x} = e \in H$. Put e, x for x, y in **S**, and we find $e \circ \hat{x} = \hat{x} \in H$; this shows H satisfies **S2**. To verify that H satisfies **S1**, take any $x, y \in H$. We know $\hat{y} \in H$, and we know also (Theorem (iii), p. 53) $(\hat{\hat{y}}) = y$. So put x, $\hat{y}$ for x, y in **S**, and get $x \circ y \in H$.

In additive and multiplicative notations, **S** becomes

$x, y \in H \Rightarrow x - y \in H,$

and

$$x, y \in H \Rightarrow xy^{-1} \in H,$$

respectively.

Example 93. If H is a subgroup of $(G, \circ)$, and $x \in H$, then all the powers x^n (n any integer) belong to H.

If $(G, \circ)$ is Abelian, so is $(H, \circ)$.

Example 94. G itself is a subgroup of any group $(G, \circ)$. So is the one-element set $\{e\}$ (check **S1** and **S2** – the only values x and y can have are $x = e$ and $y = e$!).

Example 95. Z is a subgroup of $(R, +)$ (check **S**: if $x, y \in Z$, then also $x - y \in Z$). $(Z, +)$ is the group of Example 84, p. 51. The set R^+ of all positive real numbers if a subgroup of $(R^*, .)$. (Example 86, p. 51) (check **S**: if x, y are positive, so is xy^{-1})). In Example 92, Y is not a subgroup of $(Z, +)$, because Y does not satisfy **S2**, although it does satisfy **S1**.

Example 96. Let m be any positive integer, and define $mZ = \{\ldots, -2m, -m, 0, m, 2m, \ldots\}$, the set of all multiples of m. Then mZ is a subgroup of $(Z, +)$ (if $x, y \in mZ$, clearly $x - y \in mZ$).

Example 97. The following are all the subgroups of the symmetric group $S(3)$ (see p. 241): $S(3), \{\iota, \alpha, \beta\}, \{\iota, \rho\}, \{\iota, \sigma\}, \{\iota, \tau\}, \{\iota\}$. Notice that they all contain the unit element ι.

Example 98. Let $\mathcal{S}$ be any set of subgroups of $(G, .)$. Then $J = \cap \mathcal{S}$ (see p. 12) is a subgroup. For let $x, y \in J$. This means, $x, y \in H$ for each member H of $\mathcal{S}$. Since each H is a subgroup, $xy^{-1} \in H$. But then $xy^{-1} \in H$ for each $H \in \mathcal{S}$, so $xy^{-1} \in J$. i.e. J satisfies **S**. In particular if H, K are subgroups of $(G, .)$, then so is $H \cap K$. In general $H \cup K$ is not a subgroup – take $H = \{\iota, \rho\}, K = \{\iota, \sigma\}$ in the last example.

5.3   Subgroup generated by a subset

From now on we shall use *multiplicative* notation for a general group unless we explicitly indicate the contrary; also we shall follow a standard practice and speak of 'the group G' instead of 'the group $(G, .)$'.

Let X be a non-empty subset of a group G, not necessarily a subgroup of G. By a *word in X*, we shall mean any element of G which can be expressed in the form

$$u_1^{m_1} u_2^{m_2} \ldots u_f^{m_f}, \tag{5.1}$$

where f is any positive integer, $u_1, u_2, \ldots, u_f$ are any elements of X (not necessarily distinct), and $m_1, m_2, \ldots, m_f$ are any integers.

The *set of all words in X* will be denoted gp X.

Example 99. gp X is the set of all elements of G which you can get, starting with the elements of X and then forming products and inverses any number of times. For example gp (x, y) includes $x, y, xy, yx, xyx, x^{-1}y$, $yx^{-1}, xy^{-1}, xy^{-1}x, x^{-1}y^{-1}, xyx^{-1}$ etc.

Example 100. There are infinitely many expressions of type (5.1), but of course they do not all give different elements of G. Taking $X = \{\alpha, \rho\}$ in the symmetric group $S(3)$ we soon find that every element of $S(3)$ is a word in X, and in many ways. For example $\beta = \alpha^2 = \alpha^{-1}\rho^2 = \alpha\rho\alpha^{-1}\rho^{-1} = \rho\alpha\rho = \rho^{17}\alpha^{-2}\rho^9\alpha^9$ etc.

THEOREM. *For any non-empty subset X of G, gp X is a subgroup of G.*

Proof. Let $x = u_1^{m_1} \ldots u_f^{m_f}$ and $y = v_1^{n_1} \ldots v_g^{n_g}$ be any elements of gp X $(u_1, \ldots, u_f, v_1, \ldots, v_g \in X)$. Then

$$xy^{-1} = u_1^{m_1} \ldots u_f^{m_f} v_g^{-n_g} \ldots v_1^{-n_1},$$

which is also a word in X. Hence gp X satisfies the condition **S** of p. 58.

DEFINITION. gp X is called the *subgroup of G generated by X.* If gp $X = G$, i.e. if every element of G is a word in X, we say *G is generated by X*, and that X is a *set of generators* for G.

Example 101. $S(3)$ is generated by $\{\alpha, \rho\}$. For every element of $S(3)$ can be written as a word in $\{\alpha, \rho\}$, e.g. $\iota = \alpha^0$, $\tau = \alpha\rho$, etc. Equally, this group is generated by $\{\sigma, \tau\}$, as the reader may confirm. Any group G is generated by G itself. If H is a subgroup of G, then $H = $ gp H.

Example 102. If G has a finite set X of generators, then G is said to be *finitely generated*. For example Z (i.e. $(Z, +)$) is generated by $\{1\}$, since every element of Z has the form $m1$ for some integer m, and $m1$ is a word, additively written, in the set $\{1\}$. See Example 103, below.

5.4 Cyclic groups

DEFINITION. A group G is called *cyclic* if there is an element x of G such that $G = \text{gp}\{x\}$. Any such element x is called a *generator of G*.

Each element of gp $\{x\}$ has the form (5.1) of section 5.3, where now all of $u_1, \ldots, u_f$ must be equal to x, i.e. it is a power x^m of x. So the cyclic group gp x consists of the elements

$$\ldots, x^{-2}, x^{-1}, x^0 = e, x, x^2, \ldots, \tag{5.2}$$

which need not all be distinct. Since $x^m x^n = x^{m+n} = x^n x^m$ for any integers m, n (Example 91, p. 54), *every cyclic group is Abelian*.

Example 103. In additive notation the elements (5.2) are

$$\ldots, -2x, -x, 0x = 0, x, 2x, \ldots.$$

For example the subgroup mZ of Z (Example 96) is cyclic and is generated by m. It is also generated by $-m$.

Example 104. If e is the unit element of a group G, then gp$\{e\}$ is the one-element subgroup $\{e\}$. For $e^n = e$ for any integer n.

There are two kinds of cyclic group gp $\{x\}$, corresponding to the two possibilities that the elements (5.2) are, (a) all distinct, or (b) not all distinct.

Infinite cyclic groups

If all the elements (5.2) are distinct, i.e. if $x^a \neq x^b$ whenever $a \neq b$, gp $\{x\}$ is infinite, and we say also that x is an *element of infinite order*.

Finite cyclic groups

If there exist distinct a, b such that $x^a = x^b$, we can assume $a > b$ and deduce $x^{a-b} = x^{b-b} = x^0 = e$. This shows that there exist positive integers p such that $x^p = e$; now define m to be the *smallest positive integer such that* $x^m = e$, m is called the *order of x*. Now we prove the following lemma.

LEMMA. *If m is the order of x, and if a, b are any integers, then* $x^a = x^b$ *if and only if* $a \equiv b \bmod m$.

Proof. First we observe that $x^{qm} = (x^m)^q = e^q = e$, for any integer q. So if $a \equiv b \bmod m$, there exists an integer q such that $a \equiv b + qm$, hence $x^a = x^{b+qm} = x^b \; x^{qm} = x^b e = x^b$. Conversely, suppose that $x^a = x^b$. As before we find $x^p = e$, where $p = a - b$. Let $p = qm + s$, where s is the residue of $p \bmod m$, so that s is one of $0, 1, \ldots, m-1$. Now $e = x^p = x^{qm+s} = x^{qm}x^s = x^s$. If $s > 0$, this contradicts the fact that m is the smallest positive integer such that $x^m = e$. Therefore $s = 0, p = a - b = qm$, which shows that $a \equiv b \bmod m$. This proves the lemma. Since each integer n is congruent mod m to exactly one of the integers $0, 1, \ldots, m-1$, gp $\{x\}$ is the set of m elements.

$$x^0 = e, x, x^2, \ldots, x^{m-1}. \tag{5.3}$$

Notice that the order of $x = m = |\text{gp } \{x\}|$.

Example 105. From the lemma, putting $b = 0$, we have that if x has order m and a is any integer, $x^a = e$ *if and only if a is a multiple of m*. Also notice $x^{-1} = x^{m-1}$. The element α of $S(3)$ (p. 241) has order 3, and gp$\{a\}$ $= \{i, \alpha, \alpha^2\} = \{i, \alpha, \beta\}$. The unit element of any group has order 1. Every non-zero element x of Z has infinite order, for $mx = 0$ only if $m = 0$. Table 2 (p. 242) shows a cyclic group of order 6.

In a given group G, its subgroups are 'features' which we look for. One way to get a subgroup is to take any *subset X* of G and find the subgroup gp X which it generates – for example each element x of G gives us a cyclic subgroup gp $\{x\}$. But there is another very natural source of subgroups, based on the idea of a group 'acting' on a set, which we define in the next section.

5.5 Groups acting on sets

Let A be a set, and $(G, .)$ a group written with multiplicative notation. If often happens that we can define a 'product' in which each pair (x, a) of elements, $x \in G$ and $a \in A$, combine to give an element of A, which we shall write $x * a$. This is very similar to a binary operation, and we can think of $*$ as a map $G \times A \to A$.

DEFINITION. If A is any set and $(G, .)$ any group, we say G *acts on A on the left by the product* $*$, if for each pair (x, a) of elements $x \in G$, $a \in A$ there is defined an element $x * a$ of A, in such a way that for all $x, y \in G$ and all $a \in A$ the following axioms hold:

A1 $e * a = a$,

and

A2 $x * (y * a) = (xy) * a$.

We can think of each element x of G 'acting' on the set A, in the sense that it changes $a \in A$ to $x * a$, which is also an element of A. Axiom **A1** says that the unit elements e 'acts' by leaving each element of A unchanged. Axiom **A2** says that if first y, and then x act, the combined effect is the same as the action of xy.

Sometimes it is convenient to allow G to act on A *on the right*. In that case we have a product $* : A \times G \to A$, so that for each pair (a, x) of elements $a \in A$ and $x \in G$ there is defined an element $a * x$ of A, in such a way that the following axioms hold:

A′1 $a * e = a$,

and

A′2 $(a * x) * y = a * (xy)$,

for all $a \in A$ and $x, y \in G$.

As we shall see from the examples below, various notations are used (instead of $*$) to denote special actions of a group on a set.

Example 106. Take any set A, and let $G = S(A)$ (Example 83, p. 51). For any $a \in A$, $\theta \in S(A)$ define $\theta * a$ to be $\theta(a)$, the image of a under θ. Then $S(A)$ acts on A (on the left) with this product: the axioms **A1** $(\iota_A * a = a)$ and **A2** $[\theta * (\phi * a) = (\theta\phi) * a]$ hold, by the definitions of ι_A and of $\theta\phi$, respectively.

Example 107. Conjugates. Let G be any group, and take $A = G$.

We shall make G act on itself on the right as follows: if $a, x \in G$ define $a * x = x^{-1}ax$. This element $x^{-1}ax$ is usually written a^x, and called the *conjugate*† *of a by x.* **A′1** holds, $a^e = e^{-1}ae = a$ for all $a \in G$. Also **A′2** holds, because $(a^x)^y = y^{-1}(x^{-1}ax)y = y^{-1}x^{-1}axy = (xy)^{-1}a(xy) = a^{xy}$, i.e. $(a*x)*y = a*(xy)$.

Example 108. We can make G act on the right on the set $B = \mathscr{B}(G)$ of all subsets of G, as follows. If $U \in B$, i.e. if $U \subseteq G$, and if $x \in G$, define $U*x$ to be the set $Ux = \{ux | u \in U\}$. (For example take $U = \{ı, \alpha\}$, subset of the group $S(3)$. Then $U\sigma = \{ı\sigma, \alpha\sigma\} = \{\sigma, \rho\}$, which is again a subset of $S(3)$.) The axioms now read **A′1**: $Ue = U$, and **A′2**: $(Ux)y = U(xy)$, for any subset U of G, and any elements $x, y \in G$; these are easy to verify.

Example 109. Now we shall make G act on the same set B as in the last example, but in a different way, by defining $U*x$ to be the set $U^x = \{u^x | u \in U\}$. (In the example $U = \{ı, \alpha\}$, $x = \sigma$ we should now have $U^\sigma = \{ı^\sigma, \alpha^\sigma\} = \{\sigma^{-1}ı\sigma, \sigma^{-1}\alpha\sigma\} = \{ı, \beta\}$.) The axioms now read **A′1**: $U^e = U$, and **A′2**: $(U^x)^y = U^{xy}$; these follow readily from what was proved in Example 107.

Example 110. Let H be a subgroup of G. Then H acts on G on the left, if we define, for $a \in G$ and $x \in H$, $x*a = xa$ (i.e. the ordinary product of x and a as elements of G). **A1** and **A2** follow at once from the group axioms – in particular **A2** from the associative law in G. We could make H act on G on the right if we wanted, by defining $a*x = ax$.

5.6 Stabilizers

DEFINITION. Let G be a group which acts on a set A on the left by a product $*$, and let a be a given element of A. Then the set

$$G_a = \{x \in G | x*a = a\}$$

is called the *stabilizer* of a.

G_a is the set of all elements of G which leave a unchanged, or 'stable'. We prove now a fundamental, but elementary theorem.

THEOREM. G_a *is a subgroup of* G.

Proof. We shall verify conditions **S1** and **S2** (p. 58) for G_a.

S1. If $x, y \in G_a$, it means $x*a = a$ and $y*a = a$. Hence $xy*a =$

---

†Do not confuse this with the *power* $a^n (n \in Z)$.

$x*(y*a)$ by **A2**, so that $xy*a=x*a=a$, i.e. $xy \in G_a$, as required.

S2. If $x \in G_a$, so that $x*a=a$, then also $x^{-1}*a=a$, hence $x^{-1} \in G_a$. This follows by taking $a=b$ in the

LEMMA. *If a, b are any elements of A, and x any element of G, then* $x*a=b \Leftrightarrow a=x^{-1}*b$.

Proof. From $x*a=b$ follows $x^{-1}*(x*a)=x^{-1}*b$. But $x^{-1}*(x*a)=x^{-1}x*a$ (by **A2**) $=e*a=a$ (by **A1**). This proves that $x*a=b \Rightarrow a=x^{-1}*b$; the converse is proved similarly.

Everything we have said in this section has its counterpart for the case where G acts on A *on the right*. The stabilizer G_a is defined to be the set

$$G_a = \{x \in G \mid a*x=a\},$$

and it can be proved that G_a is always a subgroup of G – the reader can prove this, by going through the proof of the theorem above, and making the necessary changes, substituting $a*x$ for $x*a$, etc. The lemma now takes the form

$$a*x=b \Leftrightarrow a=b*x^{-1}.$$

We shall return to the theory of stabilizers in the next chapter. For the moment we consider this simply as an important way of finding subgroups of groups.

Example 111. Take $G=S(n)$, the symmetric group on $A=\{1, 2, \ldots, n\}$, which acts on A as in Example 106. The stabilizer G_1 is the set of all permutations θ of A which map 1 to 1 – for example if $n=3$

G_1 is the subgroup of two elements $\iota=1$, $\rho=(23)$.

Example 112. In Example 107, G acts on G by $a*x=x^{-1}ax$. The condition $a*x=a$, i.e. $x^{-1}ax=a$, is exactly the condition that a, x commute. For $x^{-1}ax=a \Leftrightarrow x(x^{-1}ax)=xa \Leftrightarrow ax=xa$. Thus the stabilizer of a, usually written $C(a)$ or $C_G(a)$ and called the *centralizer of a in G*, is the set of all elements x of G which commute with a. For example the centralizer of α in $S(3)$ is easily found to be the subgroup $\{\iota, \alpha, \beta\}$.

Example 113. The stabilizer of a subset U in Example 108 is the set of all x

in G such that $Ux = U$, i.e. Ux and U must be the same *sets* – the order of elements does not matter. For example if $U = \{\imath, \tau\} \subseteq S(3)$, then τ belongs to the stabilizer of U. because $U\tau = \{\imath\tau, \tau\tau\} = \{\tau, \imath\}$, and this is the same set as U.

Example 114. The stabilizer of a subset U of G in the sense of Example 109 is called the *normalizer* (denoted $N(U)$ or $N_G(U)$) of U in G. It is the set of all x in G such that $U^x = U$. For example the normalizer in $S(3)$ of $U = \{\alpha, \beta\}$ is the whole group $S(3)$, because it happens that $U^\theta = U$ for all $\theta \in S(3)$, e.g. $U^\sigma = \{\alpha^\sigma, \beta^\sigma\} = \{\beta, \alpha\} = U$.

Example 115. The stabilizer of an element a of G, which is acted on by the subgroup H as in Example 110, is $\{e\}$. For the only element $x \# * H$ such that $xa = a$ is $x = e$.

Exercises for Chapter 5

1   Let X be a given set, and Y a fixed subset of X. Show that the set $\mathscr{S}$ of all subsets A of X such that $A \supseteq Y$ is closed to the opration $\cup$ on $\mathscr{B}(X)$. Is $\mathscr{S}$ closed to $\cap$?

2   Define an operation $\circ$ on $A = \{1, 2, \ldots, n\}$, in such a way that no proper subset of A is closed to $\circ$.

3   Let $(G, .)$ be any Abelian group, and let H be a non-empty subset of G which is closed to ., i.e. H satisfies **S1** but not necessarily **S2**. Show that the set $H^* = \{xy^{-1} | x \in H, y \in H\}$ is a subgroup of G.

4   Show that the set H of all non-zero integers is not a subgroup of the group $(R^*, .)$, but that it does satisfy **S1**. Find the subgroup H^* (see Exercise 3) in this case.

5   Let m, n be positive integers. Prove that $mZ \cap nZ = lZ$, where l is the least common multiple of m, n.

6   H is a subgroup of a group $(G, .)$, and x, y are elements of G. Prove that if *any two* of x, y, xy are in H, then so is the third.

7   If H, K are subgroups of a group G and if $H \cup K$ is also a subgroup, prove that either $H \subseteq K$ or $K \subseteq H$.

8   Prove that $S(3)$ is generated by $\{\alpha, \rho\}$, and also by $\{\sigma, \tau\}$.

9   If X is a subset of G such that $xy = yx$ for every pair (x, y) of elements of X, prove that gp X is Abelian.

10   Let m, n be any positive integers. Regarding these as elements

of the group Z, prove that gp $\{m, n\} = $ gp $\{d\}$, where d is the highest common factor of m and n.

11 Show that gp X is the smallest subgroup of G which contains X, i.e. if H is any subgroup of G and $H \supseteq X$, prove that $H \supseteq$ gp X.

12 Show that gp $X = \cap \mathscr{S}_X$, where $\mathscr{S}_X$ is the set of all subgroups of G which contain X.

13 Find the order of all the elements of $A(4)$ (p. 243). Also show that the permutation α of Exercise 9, p. 40, has order p.

14 If x is a generator of a cyclic group G, show that also x^{-1} generates G. Find all the generators of $(Z, +)$. Find all the generators of a cyclic group $G = $ gp $\{x\}$ of order 8.

15 If G is a *finite* group, prove that any non-empty subset H of G which satisfies **S1** (p. 58) also satisfies **S2**, i.e. is a subgroup. (Prove that every element of G has finite order.)

16 Find the centralizers of all the elements of $S(3)$.

17 Find the normalizer in $S(4)$ of the set $\{\beta, \beta^{-1}\}$, where $\beta = (1234)$.

18 If G acts on a set A on the left and if $a \in A$, $x \in G$, prove $G_{x * a} = x G_a x^{-1}$.

19 When G acts on $\mathscr{B}(G)$ as in Example 108, p. 64, show that the stabilizer of a subgroup H of G, regarding H as element of $\mathscr{B}(G)$, is H itself.

6 Cosets

6.1 The quotient sets of a subgroup

Throughout this section G is a group, written with multiplicative notation, and H is a subgroup of G. We are going to show that H determines two *partitions* of G, called the right and left *quotient sets*, respectively, of G by H. This fact is a very striking feature of group theory; there is no counterpart to it in the theory of general gruppoids, or even semigroups.

We need first some notation (already introduced in Example 108, p. 64). If U is any subset of G and x any element of G let $Ux = \{ux | u \in U\}$. This is again a subset of G; if U is a finite set $\{u_1, \ldots, u_n\}$ then $Ux = \{u_1x, \ldots, u_nx\}$. Similarly define $xU = \{xu | u \in U\}$. In general, $Ux \neq xU$. But from the associative law in G we find at once that $(Ux)y = U(xy)$, $(xU)y = x(Uy)$ and $x(yU) = (xy)U$ for any $x, y \in G$, so we need not use brackets in these products. Also $eU = Ue = U$.

DEFINITION. Let H be a subgroup of a group G. Any set Hx, with $x \in G$, is called a *right coset* of H in G. The set of all right cosets Hx, $x \in G$, is called the *right quotient set* of G by H, denoted G/H. Similarly any set xH is called a *left coset* of H in G, and the set of all left cosets xH, $x \in G$, is called the *left quotient set* of G by H, denoted $G \backslash H$.

Now let x, y be elements of G. Write $u = yx^{-1}$, which is equivalent to $y = ux$. Since y belongs to Hx if and only if $y = ux$ for some element u of H, we have

(i) LEMMA. $y \in Hx \Leftrightarrow yx^{-1} \in H$.

Next we define a relation $\sim$ on G as follows: if $x, y \in G$, '$x \sim y$' shall mean '$yx^{-1} \sim H$'.

(ii) LEMMA. $\sim$ *is an equivalence relation on* G.

Proof. **E1** holds, for if $x \in G$, then $xx^{-1} = e \in H$, because H is a subgroup. **E2** holds, for if $x \sim y$, i.e. $yx^{-1} \in H$, then also $xy^{-1} = (yx^{-1})^{-1} \in H$ by **S2**, so that $y \sim x$. Finally **E3** holds, for if $x \sim y$ and $y \sim z$, i.e. yx^{-1} and zy^{-1} both belong to H, then **S1** shows that $zx^{-1} = (zy^{-1})(yx^{-1}) \in H$, so that $x \sim z$.

Now consider the *equivalence class* E_x of a given element x of G. By definition (p. 19) this is the set of all $y \in G$ such that $x \sim y$, i.e. such that $yx^{-1} \in H$. Thus lemma (i) shows that E_x is exactly the right coset Hx, and we can apply the fundamental theorem on equivalence relations (p. 20), and obtain the

THEOREM. *The set* $G/H = \{Hx \mid x \in G\}$ *is a partition of* G. *If* x, y *are elements of* G *then* $Hx = Hy \Leftrightarrow yx^{-1} \in H$.

In exactly the same way, but using the equivalence relation defined by '$x^{-1}y \in H$', we have the analogous result for the left quotient set: $G \backslash H = \{xH \mid x \in G\}$ *is a partition of* G. *If* x, y *are elements of* G *then* $xH = yH \Leftrightarrow x^{-1}y \in H$.

Example 116. H itself is both a right and a left coset of H, because $H = He = eH$. Putting $x = e$ in the theorem above, we have that $Hy = H$ if and only if $y \in H$; similarly $yH = H$ if and only if $y \in H$. But in general the right and left cosets Hx and xH are different, and so G/H and $G \backslash H$ are different partitions of G. We give below the right and left cosets of the subgroup $H = \{\iota, \rho\}$ of the symmetric group $S(3)$ (see p. 241). There are three *right cosets:*

$$H = H\iota = H\rho = \{\iota, \rho\},$$
$$H\alpha = H\tau = \{\alpha, \rho\},$$

and

$$H\beta = H\rho = \{\beta, \tau\}.$$

So the right quotient set is $G/H = \{\{\iota, \rho\}, \{\alpha, \rho\}, \{\beta, \tau\}\}$. There are also three *left cosets:*

$$H = \iota H = \rho H = \{\iota, \rho\},$$
$$\alpha H = \rho H = \{\alpha, \tau\},$$

and

$$\beta H = \tau H = \{\beta, \rho\}.$$

So the left quotient set is $G \backslash H = \{\{\iota, \rho\}, \{\alpha, \tau\}, \{\beta, \rho\}\}$.

Example 117. In additive notation a coset Hx is written $H + x$, and xH is written $x + H$. If $(G, +)$ is an Abelian group, then of course the right and left cosets coincide.

Example 118. Let m be a fixed positive integer, and mZ the subgroup of $(Z, +)$ consisting of all multiples of m (Example 96, p. 59). Taking $G = Z$ and $H = mZ$, the equivalence relation $\sim$ above is given by the rule: if $x, y \in Z$, '$x \sim y$' means '$y - x \in mZ$', in other words, '$x \sim y$' is the same as '$x \equiv y$ mod m'. So the cosets of mZ in Z are exactly the residue classes (congruence classes) of Z mod m; the class E_x on p. 22 is the coset $x + mZ$. Therefore in this case the quotient set (we need not say 'right' or 'left', because $(Z, +)$ is Abelian) Z/mZ has m elements $r + mZ$ ($r = 0, 1, \ldots, m-1$).

6.2 Maps of quotient sets

It is at first rather difficult to realize that the notation Hx for a right coset is ambiguous, because the same coset may also be written Hy, for some y different from x. In fact the theorem of the last section tells us exactly when this happens:

(i) $Hx = Hy$ *if and only if* $yx^{-1} \in H$.

Similarly for left cosets

(ii) $xH = yH$ *if and only if* $x^{-1}y \in H$.

This is particularly important when we try to define a map θ of G/H (or of $G \backslash H$) into some set B. We usually do this by a rule of the form

$$\theta(Hx) = \hat{x}, \tag{6.1}$$

where $\hat{x}$ is some element of B, $\hat{x}$ being defined for each x of G. But if $Hx = Hy$, then rule (6.1) also gives $\theta(Hy) = \hat{y}$, so that unless $\hat{x} = \hat{y}$

the rule (6.1) is inconsistent – it does not give a well-defined image of the coset $Hx = Hy$ under θ. So whenever we define a map $\theta:G/H\to B$ by a rule of this type, we must verify that the rule is consistent: rule (6.1) *is consistent if and only if $\hat{x} = \hat{y}$ for any x, y of G such that $yx^{-1} \in H$.* A similar condition applies for maps of $G\backslash H$.

Example 119. Take $G = S(3)$ and $H = \{\iota, \rho\}$ as in Example 116. We might try to define a map $\theta:G/H\to G$ by the rule $\theta(Hx) = x^2$ for any $x \in G$. But this is not consistent, for $\alpha\rho^{-1} \in H$, i.e. $H\alpha = H\rho$, but $\alpha^2 \neq \rho^2$. Thus the image under θ of coset $\{\alpha, \rho\}$ is not well defined.

6.3 Index. Transversals

THEOREM. *Let H be any subgroup of a group G. Then $G/H \simeq G\backslash H$. In particular if either one of the two quotient sets is finite, then they both are, and they have the same order.*

DEFINITION. *If G/H is finite then $|G/H|$ is called the index of H in G.*

Proof of the theorem. We have to show that G/H and $G\backslash H$ are similar sets (p. 38), i.e. that there is a bijective map $\theta:G/H\to G\backslash H$. We attempt to define θ by the rule

$$\theta(Hx) = x^{-1}H, \qquad \text{for any } x \in G, \tag{6.2}$$

but must first verify consistency (taking $\hat{x} = x^{-1}H$). If x, $y \in G$ are such that $yx^{-1} \in H$, then also $xy^{-1}H = y^{-1}H$ by section 6.2(ii) above, so the rule (6.1) *is* consistent, and does define a map θ. To show θ is bijective, define similarly $\theta':G\backslash H\to G/H$ by the rule $\theta'(xH) = Hx^{-1}$ (and show this is consistent). It is very easy to see that $\theta'\theta = \iota_{G/H}$ and $\theta\theta' = \iota_{G\backslash H}$, and it follows that θ is bijective (section 3.6(iii), p. 34).

Example 120. If G is finite, every subgroup H has finite index, e.g. $|S(3)\backslash H| = 3$ in Example 116. Even if G is infinite it is possible for a subgroup of G to have finite index, for example $m\mathbb{Z}$ has index m in $\mathbb{Z}$ (Example 117).

Transversals

A subset X of G is called a *right transversal†* *for H in G* if each right coset of H in G contains exactly one member of X. If H has finite index n in G, then $X = \{x_1, \ldots, x_n\}$ must have n elements, one from each right coset, and then $G/H = \{Hx_1, \ldots, Hx_n\}$. Similarly we may define *left transversals*.

Example 121. There are in general many transversals of H in G. For example the sets $\{\iota, \alpha, \beta\}$, $\{\rho, \alpha, \tau\}$, $\{\rho, \sigma, \tau\}$ are all right transversals of H in $S(3)$ (Example 116). The set $X = \{0, 1, \ldots, m-1\}$ is a transversal of mZ in Z (Example 117).

6.4 Lagrange's theorem

We may notice in Example 116 (p. 69) that all the cosets of the subgroup H have the same order, i.e. they all have the same order as H (which is itself a right and a left coset). This is an example of the following general fact.

LEMMA. *Let H be a subgroup of G and x any element of G. Then $H \simeq Hx$ and $H \simeq xH$. In particular, if H is finite then*

$$|H| = |Hx| = |xH| \text{ for all } x \in G.$$

Proof. Define maps $\theta : H \to Hx$ and $\theta' : Hx \to H$ as follows: if $u \in H$ let $\theta(u) = ux$, and if $v \in Hx$ let $\theta'(v) = vx^{-1}$ (notice that if $v \in Hx$ then $vx^{-1} \in H$, by section 6.1, Lemma (i)). Verify that $\theta\theta' = \iota_{Hx}$, and $\theta'\theta = \iota_H$; this shows θ is bijective (section 3.6(iii), p. 34). Thus $H \simeq Hx$; similarly $H \simeq xH$.

Suppose now that H is a subgroup of a *finite* group G. Let $m = |H|$ and $n = |G/H|$. Then G/H is a partition of G into n cosets, each of which has m elements. Therefore the order of G is mn. This gives the following theorem, which is one of the oldest in group theory.

---

†Or a *set of right coset representatives*.

LAGRANGE'S THEOREM. *If H is a subgroup of a finite group G, then the order of H divides the order of G, and $|G| = |H|\ |G/H|$.*

We can write this formula $|G/H| = |G|/|H|$, which explains the notation and the term 'quotient set'. It is clear that we could use the set $G\backslash H$ of all *left* cosets xH, instead of G/H; this gives another version of Lagrange's theorem, namely $|G\backslash H| = |G|/|H|$. Of course, this can be deduced from the theorem in section 6.3, which shows that $|G/H| = |G\backslash H|$.

Example 122. The order (p. 62) of an element x of a finite group G divides the order of G. For the order of x is the order of the cyclic subgroup $H = \text{gp}\{x\}$ of G.

Example 123. If a group G has prime order p, then G is cyclic. For let $x \in G$, $x \neq e$. The order of the subgroup $H = \text{gp}\{x\}$ must divide p, which is prime. So $|H|$ can only be 1 or p. But H has at least two elements e and x, so in fact $|H| = p$, i.e. H is the whole of G. Therefore $G = \text{gp}\{x\}$, and so is cyclic.

6.5 Orbits and stabilizers

Suppose that the group G acts on a set A on the left, as in sections 5.5 and 5.6. This action of G defines a relation on A as follows: if $a, b \in A$, let '$a \sim b$' mean 'there is some $x \in G$ such that $x * a = b$'.

(i) LEMMA. $\sim$ *is an equivalence relation on A.*

Proof. **E1.** If $a \in A$, then $e * a = a$, hence $a \sim a$. **E2.** If $a \sim b$, so that $x * a = b$ for some $x \in G$, then $x^{-1} * b = a$ by the lemma on p. 65, hence $b \sim a$, **E3.** If $a \sim b$ and $b \sim c$, there exist elements $x, y \in G$ such that $x * a = b$ and $y * b = c$. Then $(yx) * a = y * (x * a) = y * b = c$, and this shows that $a \sim c$.

DEFINITION. Let a be a fixed element of A. Then the equivalence class

$$O_a = \{x * a \,|\, x \in G\}$$

of a under $\sim$ is called the *orbit of a under G.*

By the fundamental theorem on equivalence relations (p. 20) the set of orbits of the elements of A is a partition of A; if A is finite and $A_1, \ldots, A_n$ are the different orbits, then

$$A = A_1 \cup \ldots \cup A_n$$

and $A_i \cap A_j = \varnothing$ whenever $i \neq j$. If A is itself a single orbit, we say A is *transitive*; this means that if a, b are any elements of A, there is always at least one element x of G such that $x*a = b$.

(ii) LEMMA. *Let a be a fixed element of A and G_a its stabilizer. Then for any elements x, y of G, $x^{-1}y \in G_a \Leftrightarrow x*a = y*a$.*

Proof. Taking $b = y*a$ in the lemma of section 5.6 (p. 65), we have $x*a = y*a \Leftrightarrow a = x^{-1}*(y*a) = (x^{-1}y)*a \Leftrightarrow x^{-1}y \in G_a$. This lemma allows us to prove a theorem which connects the orbit O_a of a given element a of A, with its stabilizer G_a.

THEOREM. *For any fixed element a of A, $G \backslash G_a \simeq O_a$. In particular if G is finite, $|O_a| = |G \backslash G_a|$, hence by Lagrange's theorem*

$$|O_a| = |G|/|G_a|.$$

Proof. We may define a map $\theta: G \backslash G_a \to O_a$ by the rule $\theta(xG_a) = x*a$, for all $x \in G$. For the lemma above tells us that this rule is consistent in the sense of section 6.2. We want to prove that θ is bijective. First θ is clearly surjective, by the definition of O_a, and if x, $y \in G$ are such that $\theta(xG_a) = \theta(yG_a)$, then we have $x*a = y*a$ which implies $x^{-1}y \in G_a$, i.e. $xG_a = yG_a$, again by lemma (ii). So θ is injective, and this proves the theorem.

In the above, we assumed that the group G acts on A *on the left*. There is an exactly parallel theory for the case where G acts on A on the right. In that case we define the orbit of an element a of A under G to be the set

$$O_a = \{a*x \mid x \in G\}.$$

These orbits are the equivalences classes of A, for an equivalence relation $\sim$ on A defined as follows: if a, $b \in A$, let '$a \sim b$' mean 'there is some $x \in G$ such that $a*x = b$'. Lemma (ii) above has its counterpart, which says that for a fixed element $a \in A$ and any $x, y \in G$, then

$$yx^{-1} \in G_a \Leftrightarrow a*x = a*y.$$

Of course the stabilizer G_a here is defined as on p. 74. Finally we have a theorem which says that for a fixed element a of G, $G/G_a \simeq O_a$, and if G is finite $|O_a| = |G/G_a| = |G|/|G_a|$. Notice that, this time, we use the set G/G_a of *right* cosets of G_a. These facts are easily proved by making the appropriate changes to the proofs of their 'left-handed' versions.

It may be useful to repeat what has been proved concerning the order of an orbit, as the following theorem.

ORBIT-STABILIZER THEOREM. *If a finite group G acts on a set A (either on the right or left), then for any $a \in A$ the order of the orbit O_a under G is given by*

$$|O_a| = |G|/|G_a|,$$

when G_a is the stabilizer of a. (Notice that the order of O_a always divides the order of G.)

Example 124. Let r, n be fixed positive integers, $r \leq n$, and let B_r be the set of all subsets of order r of the set $A = \{1, 2, \ldots, n\}$. The symmetric group $S(n)$ acts on B_r on the left, defining $\theta*U$ to be the set $\theta U = \{\theta(u) | u \in U\}$, for any $U \in B_r$ and $\theta \in S(n)$. Now B_r is *transitive*, because if $U = \{u_1, \ldots, u_r\}$ and $V = \{v_1, \ldots, v_r\}$ are any subsets of A of order r, it is possible to find a permutation θ of A such that $\theta(u_1) = v_1, \ldots, \theta(u_r) = v_r$, so $\theta U = V$. So B_r is the orbit of any one of its members, e.g. of $U = \{1, 2, \ldots, r\}$. The stabilizer G_U of U is the set of all permutations

$$\theta = \begin{pmatrix} 1 & 2 & \ldots r & r+1 \ldots n \\ a_1 & a_2 \ldots a_r & a_{r+1} \ldots a_n \end{pmatrix}$$

such that $\{a_1, \ldots, a_r\} = U$; so $a_1, \ldots, a_r$ can be any of the $r!$ permutations of $1, \ldots, r$, and $a_{r+1}, \ldots, a_n$ can be any of the $(n-r)!$ permutations of $r+1, \ldots, n$. Thus $|G_U| = r! \ (n-r)!$ and we have $|B_r| = $ orbit of $U| = |S(n)|/|G_U| = n!/[r!(n-r)!]$ – this is the well-known fact that the number of subsets of r elements of a set of n elements, is equal to the binomial coefficient

$$\binom{n}{r} = n!/[r!(n-r)!].$$

Example 125. Sylow's theorem. There is no converse to Lagrange's theorem, i.e. if G is a group of finite order g, and if m is an integer dividing g, there may be no subgroup H of G of order m. However if p is a prime, and if p^a is the highest power of p which divides g, then G *has at least one subgroup of order* p^a. This theorem was proved by L. Sylow in 1873. Any subgroup H of G of order p^a is called a *Sylow p-subgroup of G.*

Let A be the set of all *subsets* of G of order p^a; we make G act on A as in Example 108, p. 64, i.e. if $U \in A$ and $x \in G$ we define $U*x$ to be the set Ux. Let $A_1, \ldots, A_n$ be the different orbits, so that

(a) $|A| = |A_1| + \cdots |A_n|.$

Each A_i is the orbit under G of some set U_i of some order p^a. If H_i is the stabilizer of U_i we have by the Orbit-Stabilizer theorem above

(b) $|A_i| = |G|/|H_i|$ $(i = 1, \ldots, n).$

If $u \in U_i$ and $x \in H_i$ then $ux \in U_i$, because $U_i x = U_i$. Keeping u fixed and letting x run over H_i we see that $uH_i \subseteq U_i$, for any $u \in U_i$. Therefore U_i is the union of all the left cosets uH_i, as u runs over U_i – of course these cosets may not all be distinct. But if the number of distinct ones is r_i, then $|U_1| = r_i|H_i|$, because distinct left cosets of H_i are disjoint, and they all have order $|H_i|$. This shows that $|H_i|$ divides $|U_i| = p^a$, and since the only factors of p^a are smaller powers of p,

(c) $|H_i| = p^{a_i}$,   for some $a_i \le a (i = 1, \ldots, n).$

We can write $g = |G| = kp^a$, where p does not divide k. From (b) and (c) follow

(d) $|A_i| = kp^{d_i}$,   where $d_i = a - a_i \ge 0$ $(i = 1, \ldots, n).$

By the last example,

$$|A| = \binom{kp^a}{p^a}.$$

We saw in Example 42, p. 23, that this $\equiv k \bmod p$. But p does not divide k, so p does not divide $|A|$, and then by (a) and (d), p does not divide the integer

$$k(p^{d_1} + \cdots + p^{d_n}).$$

If $d_i > 0$ for all $i = 1, \ldots, n$, then p would divide this integer. So there must be some i such that $d_i = 0$. Then for this i, $a_i = a$, i.e. $|H_i| = p^a$. This proves Sylow's theorem, because H_i is a subgroup of G of order p^a.†

As an example, the order of the group $A(4)$ (p. 243) is $12 = 2^2 \cdot 3$. Sylow's theorem tells us that $A(4)$ must have at least one subgroup of order $2^2 = 4$, and at least one of order 3. In fact, $A(4)$ has exactly one Sylow two-subgroup V, and four Sylow three-subgroups $\{e, a, p\}$, $\{e, b, s\}$, $\{e, c, q\}$, $\{e, d, r\}$. Notice that $A(4)$ has no subgroup of order 6, although 6 divides $12 = |A(4)|$.

6.6   Conjugacy classes. Centre of a group

We can make any group G act on itself, on the right, by the rule $a*x = x^{-1}ax = a^x$ (see Example 107, p. 63). The orbit O_a of a given element $a \in G$ under this action, is called the *conjugacy class* of a. According to the general theory of the last section, O_a is the equivalence class of a, with respect to the equivalence relation $\sim$ on G defined by: $a \sim b$ if there exists some $x \in G$ such that $x^{-1}ax = b$. Elements a, b of G which are related in this way, are said to be *conjugate* in G. Thus the conjugacy class of a, is the set of elements $x^{-1}ax$, for all $x \in G$.

From the general theory of orbits, we know that the set of all the conjugacy classes in G is a partition of G. So if G is finite, and if $C_1, \ldots, C_n$ are the different conjugacy classes of G, we have

$$G = C_1 \cup \ldots \cup C_n, \tag{6.3}$$

and $C_i \cap C_j = \varnothing$ whenever $i \neq j$. It is usual to arrange the notation so that $C_1 = \{1\}$, the conjugacy class containing 1 (notice that $x^{-1}1x = 1$ for all $x \in G$, so that the only element conjugate to 1, is itself). Counting the elements on both sides of (6.3), we get

$$|G| = |C_1| + \cdots + |C_n|. \tag{6.4}$$

This is called the *class equation* for G.

Suppose that g_i is an element of the class C_i. Then C_i is the orbit of g_i under the action $a*x = x^{-1}ax$ described at the beginning of this section. The stabilizer G_{g_i} is the set of all $x \in G$ such that

---

†This proof of Sylow's theorem is due to H. Wielandt (1959).

$x^{-1}g_ix = g_i$, i.e. such that $g_ix = xg_i$; this set is called the *centralizer* of g_i in G, and is denoted $C_G(g_i)$ (see Example 112, p. 65). $C_G(g_i)$ is a subgroup of G, because it is the stabilizer G_{g_i} of g_i, and a stabilizer is always a subgroup (see section 5.6). Moreover the Orbit-Stabilizer theorem gives a useful formula for the order of class C_i, namely

$$|C_i| = |G|/|C_G(g_i)|, \qquad \text{for any } g_i \in C_i. \tag{6.5}$$

It follows from equation (6.5) that (for a finite group G) *the order of every conjugacy class of G, divides the order $|G|$ of G.*

Example 126. The conjugacy classes of $G = S(3)$ are $C_1 = \{\iota\}$, $C_2 = \{\rho, \sigma, \tau\}$, $C_3 = \{\alpha, \beta\}$. These can be calculated, using the multiplication table on p. 241. For example the class C_3 containing α consists of the elements $x^{-1}\alpha x$, for all $x \in G$. Taking $x = \iota, \alpha, \beta, \rho, \sigma, \tau$ in turn, we find $x^{-1}\alpha x = \alpha, \alpha, \alpha, \beta, \beta, \beta$; thus $C_3 = \{\alpha, \beta\}$. We also see from this calculation that $C_G(\alpha) = \{\iota, \alpha, \beta\}$, so that $|C_3| = 2 = |G|/|C_G(\alpha)|$, in agreement with equation (6.5) above.

DEFINITION. The *centre* of a group G is the set $Z(G)$ of all elements $x \in G$ which commute with every element of G.

Thus an element $x \in G$ belongs to $Z(G)$, if and only if x belongs to $C_G(g)$, for *all* $g \in G$; this shows that

$$Z(G) = \cap \mathscr{S},$$

where $\mathscr{S}$ is the set $\{C_G(g) | g \in G\}$. The intersection of any set $\mathscr{S}$ of subgroups of G is itself a subgroup (see Example 98, p. 59), so it follows that $Z(G)$ *is a subgroup of G.* (This could also be proved directly, using the conditions **S1, S2** of section 5.2.)

Example 127. $Z(G) = G$ if and only if G is Abelian.

Example 128. Let g be an element of a group G. Then $g \in Z(G)$ if and only if the conjugacy class of g is the one-element set $\{g\}$. For $g \in Z(G) \Leftrightarrow x^{-1}gx = g$, all $x \in G \Leftrightarrow$ conjugacy class of g is $\{g\}$. It follows that the centre of $S(3)$ is the group $\{\iota\}$, since the only one-element conjugacy class of $S(3)$ is $C_1 = \{\iota\}$ (see Example 126). (We shall find in Example 131, below, that $Z[S(n)] = \{\iota\}$, for all $n \geq 3$.)

Example 129. Suppose p is a prime number. A finite group p is called a *p-group* if its order $|G|$ is some power p^a of p; here a can be any integer $a \geq 0$. Any subgroup H of a p-group G is also a p-group, since by Lagrange's theorem (p. 73) the order of H must divide $|G| = p^a$, and the only divisors of p^a are powers p^b of p (with $b \leq a$). Similarly, the orders of the conjugacy classes $C_1, \ldots, C_n$ of a p-group G are all powers of p, because these orders divide $|G|$ (see equation (6.5) above). So if G is a p-group of order $p^a > 1$, its conjugacy classes $C_1, \ldots, C_n$ have orders $p^{a_1}, \ldots, p^{a_n}$, where $a_1, \ldots, a_n$ are integers ≥ 0; thus the class-equation (6.4) reads

$$p^a = p^{a_1} + p^{a_2} + \cdots p^{a_n}.$$

If C_1 is the class $\{1\}$, then of course $p^{a_1} = 1$, i.e. $a_1 = 0$. Now suppose there are no other one-element classes C_i except C_1; then $a_i \geq 1$ for all $i = 2, \ldots, n$. Then p divides $p^{a_2} + \ldots + p^{a_n}$, and also p divides p^a, so from the equation above we find that p divides $p^a - (p^{a_2} + \cdots + p^{a_n}) = p^{a_1} = 1$. This is not true, hence there must be some one-element classes C_i, apart from C_1. But this means that $Z(G)$ contains elements other than 1 (see Example 128). We have proved the Theorem: *If G is a p-group of order $|G| > 1$, then its centre $Z(G)$ also has order $|Z(G)| > 1$.* This theorem does *not* hold for arbitrary finite groups G. For example $G = S(3)$ has order > 1, but $|Z(G)| = 1$ (Example 128).

Example 130. Conjugacy classes of the symmetric groups. Let n be a positive integer, and let π, α be any elements of the symmetric group $S(n)$. There is a very easy rule for calculating the element $\pi\alpha\pi^{-1}$, as follows. First write α in the *cycle notation* of section 3.7; for example if $n = 8$, we might have

$$\alpha = (a_1, a_2, a_3)(b_1, b_2)(c_1, c_2)(d_1), \tag{6.6}$$

where $a_1, a_2, \ldots, d_1$ are the integers $1, 2, \ldots, 8$ in some order. Then *a cycle-expression for $\pi\alpha\pi^{-1}$ is obtained by applying π to every element in the cycle-expression for α*; in our example, we get

$$\pi\alpha\pi^{-1} = (\pi a_1, \pi a_2, \pi a_3)(\pi b_1, \pi b_2)(\pi c_1, \pi c_2)(\pi d_1), \tag{6.7}$$

where we have written $\pi a_1, \pi a_2, \ldots$ instead of $\pi(a_1), \pi(a_2), \ldots$ (to avoid a forest of brackets!). To prove equation (6.7), notice that $\pi\alpha\pi^{-1}$ maps πa_1 to πa_2, since π^{-1} maps πa_1 to a_1, then α maps a_1 to a_2, then π maps a_2 to πa_2. Similarly $\pi\alpha\pi^{-1}$ maps πa_2 to πa_3, and πa_3 to πa_1; hence $\pi\alpha\pi^{-1}$ contains a cycle $(\pi a_1, \pi a_2, \pi a_3)$. In the same way, we find that $\pi\alpha\pi^{-1}$ contains all the other cycles shown in equation (6.7), which is therefore a cycle-expression for $\pi\alpha\pi^{-1}$.

We define the *cycle-pattern* of an element α of $S(n)$ to be the symbol

$[1^{r_1}2^{r_2} \ldots n^{r_n}]$, if a cycle-expression for α has r_1 1-cycles, r_2 2-cycles, etc. For example, the element α of $S(8)$ given by equation (6.6), has cycle-pattern $[1^1 2^2 3^1]$; moreover we see from equation (6.7) that $\pi\alpha\pi^{-1}$ has this same cycle-pattern, for any π in $S(8)$; this means that all the elements β in the conjugacy class of α, have the same cycle-pattern as α. Conversely, suppose that β is any element of $S(8)$ having this cycle-pattern $[1^1 2^2 3^1]$. then $\beta = (a'_1, a'_2, a'_3)(b'_1, b'_2)(c'_1, c'_2)(d'_1)$, the elements $a'_1, a'_2, \ldots, d'_1$ being the integers $1, 2, \ldots, 8$ in some order. But then there exists an element π of $S(8)$, namely

$$\pi = \begin{pmatrix} a_1, a_2, a_3, b_1, b_2, c_1, c_2, d_1 \\ a'_1, a'_2, a'_3, b'_1, b'_2, c'_1, c'_2, d'_1 \end{pmatrix}$$

such that $\pi\alpha\pi^{-1} = \beta$, as we see by using this π in equation (6.7). Thus β is conjugate to α. This argument works generally, and gives the following theorem.

THEOREM. *Elements α, β of the symmetric group $S(n)$ are conjugate in $S(n)$, if and only if they have the same cycle-pattern $[1^{r_1}2^{r_2} \ldots n^{r_n}]$.*

The cycle-pattern $[1^{r_1}2^{r_2} \ldots n^{r_n}]$ of an element α of $S(n)$ must satisfy the equation

$$r_1 + 2r_2 + 3r_3 + \cdots + nr_n = n, \tag{6.8}$$

since, in a cycle expression for α, the elements of $\{1, \ldots, n\}$ all occur; r_1 of them occur in 1-cycles,[†] $2r_2$ of them in 2-cycles, and so on. This remark, together with the theorem above, tells us how to find all the conjugacy classes of $S(n)$: there is one such class for each cycle-pattern satisfying equation (6.8), and it consists of all elements of $S(n)$ having this cycle-pattern. For example there are three cycle-patterns satisfying equation (6.8) with $n=3$, viz. $[1^3 2^0 3^0]$, $[1^1 2^1 3^0]$ and $[1^0 2^0 3^1]$. It is usual to simplify the notation for cycle-patterns, by omitting terms i^{r_i} with $r_i = 0$, and writing i^1 simply as i. With these conventions, the three patterns above appear as $[1^3]$, $[12]$, $[3]$. Each of these gives a conjugacy class of $S(3)$ – these are the classes C_1, C_2, C_3 (respectively) found in Example 126. For example, C_2 consists of all elements of $S(3)$ with cycle-pattern $[12]$, i.e. $\rho = (1)(23), \sigma = (2)(13)$ and $\tau = (3)(12)$. Notice that $\iota = (1)(2)(3)$ is only element with cycle-pattern $[1^3]$; this gives the class $C_1 = \{\iota\}$.

[†]To determine the cycle-pattern of a permutation α, it is essential to count all the one-cycles, even though these are usually omitted when α is written in cycle notation (see p. 37).

Example 131. To find the centre $Z[S(n)]$ of the symmetric group $S(n)$, may use the rule for calculating $\pi\alpha\pi^{-1}$ given in the last example. If $n = 1$, $n = 2$, then $S(n)$ is Abelian, and so is equal to its centre (see Example 127). Assume now that $n \geq 3$, and that π is any element of $Z[S(n)]$. Then $\pi\alpha\pi^{-1} = \alpha$, for all α in $S(n)$. First take† $\alpha = (12)$. By the rule mentioned (see equation (6.7)) $\pi\alpha\pi^{-1} = (\pi 1, \pi 2)$ (we write $\pi 1$ for $\pi(1)$, etc., for ease of notation). Remembering that α has *two* possible cycle-expressions, *viz.* $\alpha = (12)$ and $\alpha = (21)$, we deduce from the equation $\pi\alpha\pi^{-1} = \alpha$ that *either* (A) $\pi 1 = 1, \pi 2 = 2$, *or* (A') $\pi 1 = 2, \pi 2 = 1$. Now take $\alpha = (23)$. We have again $\pi\alpha\pi^{-1} = \alpha$, and deduce that *either* (B) $\pi 2 = 2, \pi 3 = 3$, *or* (B') $\pi 2 = 3, \pi 3 = 2$. Now it is clear that condition (A') is compatible with neither (B) nor (B'), since (A') requires $\pi 2 = 1$. But either (B) or (B') must hold, hence (A') cannot hold. In a similar way, we see that (B') is compatible with neither (A) nor (A'), since it requires $\pi 2 = 3$. Therefore (B') does not hold. It follows that we must have (A) *and* (B), i.e. $\pi 1 = 1, \pi 2 = 2, \pi 3 = 3$. Continuing in this way, using the equations $\pi\alpha\pi^{-1} = \alpha$, with $\alpha = (34), (45), \ldots, (n-1, n)$ in turn, we find $\pi i = i$ must hold for all $i = 1, 2, \ldots, n$. But this means that π is the identity map ι on $\{1, \ldots, n\}$. Since π was an arbitrary element of $Z[S(n)]$, we have proved: if $n \geq 3$, then $Z[S(n)] = \{\iota\}$.

6.7 Normal subgroups

DEFINITION. Let H be a subgroup of a group G. Then we say that H *is normal in* G, and write $H \trianglelefteq G$, if

N $\qquad Hx = xH \qquad$ for all $x \in G$.

Thus a subgroup H is normal in G if the two quotient sets G/H, $G \backslash H$ coincide. Every subgroup H of an Abelian group G is normal in G. If we multiply **N** on the left by x^{-1} we find an equivalent condition: *H is normal in G if and only if*

N' $\qquad x^{-1}Hx = H \qquad$ for all $x \in G$.

A third, and probably the most useful, form of this condition is: *A subgroup H of G is normal in G if and only if*

N'' $\qquad u \in H, x \in G \Rightarrow x^{-1}ux \in H$.

---

†$\alpha = (12)$ is short for the full cycle-expression $\alpha = (12)(3)(4) \ldots (n)$. Similarly $(\pi 1, \pi 2)$ is short for $(\pi 1, \pi 2)(\pi 3)(\pi 4) \ldots (\pi n)$. However, we lose nothing if we keep to the convention that one-cycles are omitted from cycle-expressions.

In the terminology of section 6.6, H is normal in G if H contains, with any of its elements u, also all the elements conjugate to u in G. To prove that $\mathbf{N}'$, $\mathbf{N}''$ are equivalent, observe that $x^{-1}Hx = \{x^{-1}ux | u \in H\}$, so $\mathbf{N}''$ is itself equivalent to

$\mathbf{N}'''$ $x^{-1}Hx \subseteq H$ for all $x \in G$.

Clearly $\mathbf{N}'$ implies $\mathbf{N}'''$. But if $\mathbf{N}'''$ holds, put x^{-1} in place of x and we have $xHx^{-1} \subseteq H$; multiplying on the left by x^{-1} and on the right by x gives $H \subseteq x^{-1}Hx$. With $x^{-1}Hx \subseteq H$, this gives $\mathbf{N}'$.

Example 132. $H = \{\iota, \rho\}$ is not normal in $S(3)$; for example $\rho \in H$ but $\alpha^{-1}\rho\alpha = \tau \notin H$. The subgroup $\{\iota, \alpha, \beta\}$ is normal in $S(3)$.

Example 133. In any group G, the subgroups G and $\{e\}$ are always normal. A group G, which is not a one-element group, and which has no normal subgroups except G and $\{e\}$, is called *simple*. *Any group G of prime order p is simple*, because by Lagrange's theorem *any* subgroup H of G must have order 1 or p, i.e. $H = \{e\}$ or G.

Example 134. Conjugate subgroups. In the notation of Example 107 we write $a^x = x^{-1}ax$, for any elements a, x of a group G. We prove now two identities (i) $a^x b^x = (ab)^x$, and (ii) $(a^x)^{-1} = (a^{-1})^x$, for all a, b, $x \in G$. For (i), $a^x b^x = x^{-1}axx^{-1}bx = x^{-1}abx = (ab)^x$. For (ii), $(a^x)^{-1} = (x^{-1}ax)^{-1} = x^{-1}a^{-1}x = (a^{-1})^x$. With these we can show that if H is any subgroup of G (not necessarily normal), and $x \in G$, then $H^x = x^{-1}Hx$ is a subgroup of G. For take any elements $a^x, b^x \in H^x$), then $ab, a^{-1} \in H$ because H is a subgroup, hence by (i) and (ii) $a^x b^x$ and $(a^x)^{-1} \in H^x$; this proves H^x is a subgroup. H^x is called a *conjugate subgroup* to H, in G. If H is *normal* in G, it means that H coincides with all its conjugate subgroups in G.

6.8   Quotient groups

Let G be a group, and H any normal subgroup of G. We shall now define a binary operation $\circ$ on the quotient set $G/H = G \backslash H$ by the following rule: for any cosets Hx, $Hy(x, y \in G)$ let

$$Hx \circ Hy = Hxy. \tag{6.9}$$

Our first problem is to show that this rule is consistent, i.e. if $x, x', y, y' \in G$ are such that $Hx = Hx'$ and $Hy = Hy'$, we must prove that $Hxy = Hx'y'$. This is easy, but uses the fact that H is *normal* in G, i.e. $Hz = zH$ for any $z \in G$. We have $Hxy = Hx'y$ (since $Hx = Hx') = x'Hy$ (since $Hx' = x'H) = x'Hy'$ (since $Hy = Hy') = Hx'y'$ (since $Hx' = x'H$). If H is not normal, then rule (6.9) does not work (see Example 135, below).

Example 135. Take $H = \{\iota, \rho\}$, which is not normal in $S(3)$. If we try to define $\circ$ on $S(3)/H$ by rule (6.9), we get contradictions. For example, if $A = H\alpha = H\sigma$ and $B = H\beta = H\tau$, then $A \circ B = H\alpha \circ H\beta = H\alpha\beta = H\iota = H$. But rule (6.9) also gives, $A \circ B = H\sigma \circ H\tau = H\sigma\tau = H\alpha = A$. So no such operation $\circ$ exists.

THEOREM. *Let H be a normal subgroup of a group G, and let $\circ$ be the operation defined by rule (6.9). Then $(G/H, \circ)$ is a group.*

DEFINITION. $(G/H, \circ)$ is called the *quotient group*† of G by H.

Notation

It is common to use the same notation (additive or multiplicative) for G/H as for G. So if G is written multiplicatively, we write $HxHy$ instead of $Hx \circ Hy$, and if G is written additively, we write $(H + x) + (H + y)$ instead of $(H + x) \circ (H + y)$. Thus the operation in the quotient group G/H is defined by

$$HxHy = Hxy, \qquad \text{or} \qquad (H + x) + (H + y) = H + (x + y),$$

respectively.

Proof of the theorem. Using multiplicative notation, we verify **G1**. $(HxHy)Hz = (Hxy)Hz = H(xy)z = Hx(yz) = HxHyz = Hx(HyHz)$ for all $x, y, z \in G$. **G2**. The coset $H = He$ is a unit element for G/H, for $HeHx = Hx = HxHe$ for all $x \in G$. **G3**. Hx^{-1} is the inverse of Hx, for any $x \in G$. For $HxHx^{-1} = Hxx^{-1} = He = H$, similarly $Hx^{-1}Hx = H$.

Example 136. The subgroup mZ of $(Z, +)$ (Examples 96, 117) is normal, because $(Z, +)$ is Abelian and so all its subgroups are normal. The rule for

---

† Or *factor group.*

adding two cosets is $(mZ+x)+(mZ+y)=mZ+(x+y)$. Write $E_x=mZ+x$, for short. The following table gives the addition for the group $Z/3Z$, which has three elements E_0, E_1, E_2. For example $E_2+E_2=E_4=E_1$, $E_2+E_1=E_3=E_0$, etc.

	E_0	E_1	E_2
E_0	E_0	E_1	E_2
E_1	E_1	E_2	E_0
E_2	E_2	E_0	E_1

Example 137. The table **4** on p. 244 is of a subgroup $A(4)$ of the symmetric group $S(4)$ – this is described in Example 152, p. 93. The set $V=\{1, t, u, v\}$ is easily shown to be a normal subgroup of $A(4)$. Its index, by Lagrange's theorem, is $|A(4)|/|V|=3$, and the three cosets of V in $A(4)$ are V, $A=Va$, $P=Vp$. The table below gives the multiplication for the quotient group $A(4)/V$. For example $AP=VaVp=Vap=Ve=V$, etc.

	V	A	P
V	V	A	P
A	A	P	V
P	P	V	A

Notice that $A(4)/V$ is Abelian, although $A(4)$ is not.

Exercises for Chapter 6

1 Show that $H=\{e, a, p\}$ is a subgroup of the group $A(4)$ of p. 243, and find all its right and left cosets in $A(4)$.

2 What are the cosets of the subgroup $\{e\}$ in a group G? What are the cosets of the subgroup G of G?

3 If H, K are subgroups of G and $x \in G$, prove $xH \cap xK = x(H \cap K)$.

4 If X is a right transversal of H in G, show that $Y=\{x^{-1} | x \in X\}$ is a left transversal.

5 If H is a subgroup of index n in G, show that a subset

$\{x_1, \ldots, x_n\}$ of G is a right transversal of H in G if and only if $x_i x_j^{-1} \notin H$, for all $i, j \in \{1, 2, \ldots, n\}$ such that $i \neq j$.

6 If H is a subgroup of K, and K is a subgroup of G, and if G/K and K/H are both finite, prove that $|G/H| = |G/K|\,|K/H|$.

7 If a subgroup H of G acts on G by the rule $x * a = xa$, for $x \in H$, $a \in G$ (see Example 110, p. 64), prove that the orbits are the right cosets of H.

8 H is a subgroup of G. Prove that G acts on G/H on the right if we define, for any $Hx \in G/H$ and any $z \in G$, $(Hx) * z = Hxz$. Show that G/H is transitive (p. 74) and that the stabilizer of Hx is $x^{-1}Hx$. Hence prove that $x^{-1}Hx$ is a subgroup of G, and that if H has finite index, then so has $x^{-1}Hx$, and these indices are equal.

9 Prove that $S(n)$ acts on the left on the set of A of all polynomials in n variables $x_1, \ldots, x_n$ if we define, for $f(x_1, \ldots, x_n) \in A$ and $\theta \in S(n)$, θf to be the polynomial given by

$$(\theta f)(x_1, \ldots, x_n) = f(x_{\theta(1)}, \ldots, x_{\theta(n)}).$$

Find the orbit of $f = x_1 x_2 + x_3 x_4$ under $S(4)$. Find also the stabilizer H of f and prove that H is a Sylow two-subgroup of $S(4)$.

10 Find the conjugacy classes of the groups $D(8)$ and $Q(8)$ (pp. 244, 245), and find also their centres. Notice that these are both two-groups, as defined in Example 129 (p. 79).

11 Find the possible cycle-patterns of elements of $S(4)$ (see Example 130, p. 79). Hence find the conjugacy classes of $S(4)$.

12 Find the centralizer $C_G(\alpha)$ in $G = S(n)$, of the element $\alpha = (12 \ldots n)$ of $S(n)$. [$C_G(\alpha)$ consists of all $\pi \in S(n)$ which commute with α, i.e. which satisfy $\pi \alpha \pi^{-1} = \alpha$. Use the rule in Example 130 to calculate $\pi \alpha \pi^{-1}$. Remember α has n different cycle-expressions $(123 \ldots n)$, $(23 \ldots n1)$, etc.]

13 Use the result of the last Exercise to find the order of the conjugacy class in $S(n)$ of $\alpha = (12 \ldots n)$.

14 Find the conjugacy classes of $A(4)$ (use table on p. 243). Show there is no element π of $A(4)$ such that $\pi . (123) . \pi^{-1} = (132)$, but that there is such an element π in $S(4)$.

15 Prove that if an element x of a group G commutes with elements $g_1, \ldots, g_n \in G$, then x commutes with every element

of $gp\{g_1, \ldots, g_n\}$ (see section 5.3). Hence show that if x commutes with all the elements of a set of generators of G, then x belongs to the centre $Z(G)$ of G.

16 Prove that $V = \{1, t, u, v\}$ is normal in $A(4)$, that $K = \{1, t\}$ is normal in V, but that K is not normal in $A(4)$.

17 Let a, b be real numbers, and define $\theta_{a,b}: R \to R$ by $\theta_{a,b}(x) = ax + b (x \in R)$. The set G of all such $\theta_{a,b}$ with $a \neq 0$ is a group, with product of maps as its binary operation (see Exercise 15, Chapter 4). Prove that $H = \{\theta_{1,t} | t \in R\}$ is a normal subgroup of G.

18 If U is a subgroup of G, then the *normalizer* $N_G(U)$ of U in G is defined to be the set of all $x \in G$ such that $U^x = U$ (see Example 114, p. 66. Recall $U^x = x^{-1}Ux$). It is a subgroup of G (because it is a stabilizer, see Example 114). Prove that $U \unlhd N_G(U)$.

19 Find the normalizer in $G = S(5)$ of the subgroup $U = \{\iota, \alpha, \alpha^2\}$, where $\alpha = (123)$. [$N_G(U)$ consists of all $\pi \in S(5)$ such that $\pi U \pi^{-1} = U$. Show that π belongs to $N_G(U)$ if and only if $\pi \alpha \pi^{-1} = \alpha$ or α^2. Use rule in Example 130 to find which π satisfy this condition.]

20 If G acts on A on the left and if H is the stabilizer of an element $a \in A$, show that $H \unlhd G$ if and only if H is the stabilizer of every element of the orbit O_a (see Exercise 18, Chapter 5).

21 If $H \unlhd G$ and $x \in G$ prove that $(Hx)^n = Hx^n$ for every integer n. Show that $(Hx)^n$ is the unit element of G/H if and only if $x^n \in H$.

22 Z is a subgroup of the group $(R, +)$ (Example 85, p. 51), and it is normal because $(R, +)$ is Abelian. Show that an element $x + Z(x \in R)$ has finite order in the group R/Z if and only if x is *rational*, i.e. if $x = m/n$ for some integers m, n.

7 Homomorphisms

7.1 Homomorphisms

Our study of groups has been confined until now to things, such as subgroups, quotient sets and quotient groups, which are related to a single given group G. In this chapter we consider how to *compare* two groups G and H. This is done by studying those maps of G into H, called *homomorphisms*, which 'transform' the group operation of G into that of H.

DEFINITION. Let $(G, \circ)$ and $(H, \times)$ be groups, with group operations $\circ$ and $\times$ respectively. Then a *homomorphism* of G into H is a map $\theta: G \to H$ which satisfies the condition

$$\mathbf{H} \qquad \theta(x \circ y) = \theta(x) \times \theta(y)$$

for all $x, y \in G$.

In other words, a homomorphism is a map with the property that it maps the product of x and y to the product of $\theta(x)$ and $\theta(y)$, for all x, y in G. Special names are used for special kinds of homomorphisms: a homomorphism $\theta: G \to H$ is called a *monomorphism* if θ is injective, an *epimorphism* if θ is surjective, and an *isomorphism* if t is bijective. A homomorphism $\theta: G \to G$ of a group into itself is called an *endomorphism*, or an *automorphism* if θ is also bijective; i.e. an automorphism is an isomorphism of G into G.

Example 138. The map $\phi: R \to R^+$ defined by $\phi(x) = e^x$ (Example 46, p. 27) is a homomorphism of $(R, +)$ into $(R^+, .)$, for $\phi(x+y) = e^{x+y} = e^x e^y = \phi(x)\phi(y)$, for all $x, y \in R$. Since ϕ is also bijective (Example 48), ϕ is an *isomorphism*.

Example 139. Let $(G, .)$ be a group and x an element of G. Define $\theta_x: Z \to G$ by the rule: if $n \in Z$ let $\theta_x(n) = x^n$. Then θ_x is a homomorphism of $(Z, +)$

into G. For if m, $n \in Z$ we have $\theta_x(m+n) = x^{m+n} = x^m x^n$ (Example 91, p. 54) $= \theta_x(m)\theta_x(n)$.

Example 140. If G is any group, the identity map ι_G is always a homomorphism, so in fact it is an automorphism of G.

Example 141. Let $(G, .)$ be a group and x an element of G. Define $\pi_x : G \to G$ by the rule: if $a \in G$ let $\pi_x(a) = a^x = x^{-1}ax$. Then π_x is a homomorphism, by the identity $(ab)^x = a^x b^x$ (Example 134, p. 82). Now if x, $y \in G$ we have $(a^x)^y = a^{xy}$ for all $a \in G$ (Example 107, p. 63), i.e. $\pi_y \pi_x = \pi_{xy}$. But also $a^e = a$ for all $a \in G$, which shows that $\pi_e = \iota_G$. Thus for any x, the mapping $\pi_{x^{-1}} = \theta'$, say, satisfies $\pi_x \theta' = \theta' \pi_x = \iota_G$. This proves that π_x is *bijective* (section 3.6 (iii), p. 34), therefore π_x is an automorphism of G; these automorphisms are called *inner automorphisms*. We can prove incidentally, that the set $I(G)$ of all inner automorphisms π_x, $x \in G$, is a subgroup of the symmetric group $S(G)$ (Example 83, p. 51). For if π_y, $\pi_x \in I(G)$, then $\pi_y \pi_x = \pi_{xy} \in I(G)$ and also $(\pi_x)^{-1} = \theta' = \pi_{x^{-1}} \in I(G)$.

7.2 Some lemmas on homomorphisms

If G, H are groups both written multiplicatively, then condition **H** for a map $\theta : G \to H$ to be a homomorphism reads

H $\qquad \theta(xy) = \theta(x)\theta(y)$,

for all x, $y \in G$. Suppose now that $\theta : G \to H$ is a homomorphism.

(i) LEMMA. *If x, $y \in G$ then $\theta(yx^{-1}) = \theta(y)\theta(x)^{-1}$.*

Proof. Write $u = \theta(x)$, $v = \theta(y)$ and $w = \theta(yx^{-1})$. Then by **H**, $wu = v$, so $w = vu^{-1}$, as required. If we take $x = y = e_G$, the unit element of G, then $\theta(y)\theta(x)^{-1} = \theta(e_G)\theta(e_G)^{-1} = e_H$, the unit element of H, and this proves the next lemma.

(ii) LEMMA. $\theta(e_G) = e_H$, *for any homomorphism $\theta : G \to H$.*

Now put $y = e_G$ in (i). Using (ii), we can prove the following.

(iii) LEMMA. $\theta(x^{-1}) = \theta(x)^{-1}$, *for any $x \in G$.*

Suppose now that G, H, K are groups.

(iv) LEMMA. *If $\theta: G \to H$ and $\theta: H \to K$ are both homomorphisms, then $\phi\theta: G \to K$ is a homomorphism. If θ and ϕ are both isomorphisms, then $\phi\theta$ is an isomorphism.*

Proof. We show first that $\phi\theta$ satisfies **H**. If x, $y \in G$ then $(\phi\theta)(xy) = \phi[\theta(xy)]$ (using definition of product map $\phi\theta$) $= \phi[\theta(x)\theta(y)]$ (using **H** for θ) $= \phi[\theta(x)]\phi[\theta(y)]$ (using **H** for ϕ) $= (\phi\theta)(x)(\phi\theta)(y)$ (definition of $\phi\theta$, used twice). Hence $\phi\theta$ is a homomorphism. If θ, ϕ are both isomorphisms, they are both bijective. Hence $\phi\theta$ is bijective (section 3.6(ii)); and since $\phi\theta$ is also a homomorphism, it is an isomorphism.

(v) LEMMA. *If $\theta: G \to H$ is an isomorphism, then $\theta^{-1}: H \to G$ is an isomorphism.*

Proof. We know that θ^{-1} is bijective, so we have only to verify that θ^{-1} is a homomorphism, i.e. that θ^{-1} satisfies **H**. Let u, v be any elements of H. Put $x = \theta^{-1}(u)$, $y = \theta^{-1}(v)$, so that $u = \theta(x)$ and $v = \theta(y)$. Since θ satisfies **H**, $uv = \theta(x)\theta(y) = \theta(xy)$, i.e. $\theta^{-1}(uv) = xy = \theta^{-1}(u)\theta^{-1}(v)$, as required.

Example 142. If G, H are both written additively, **H** reads $\theta(x+y) = \theta(x) + \theta(y)$, for all x, $y \in G$. Lemma (i) reads $\theta(y-x) = \theta(y) - \theta(x)$.

Example 143. If $\theta: G \to H$ is a homomorphism and G, H both written in multiplicative notation, then $\theta(x_1 \ldots x_n) = \theta(x_1) \ldots \theta(x_n)$ for any $x_1, \ldots, x_n \in G$, by applying **H** repeatedly. Taking $x_1, \ldots, x_n$ all equal to x gives $\theta(x^n) = \theta(x)^n$ for any positive integer n; lemmas (ii) and (iii) extend this to any integer n.

Example 144. The set $A(G)$ of all automorphisms of a group G is a subgroup of the symmetric group $S(G)$ (cf. Example 141). For if θ, $\phi \in A(G)$, i.e. if θ, ϕ are isomorphisms of G into G, then by lemmas (iv) and (v) $\phi\theta \in A(G)$ and $\theta^{-1} \in A(G)$. The group $I(G)$ of Example 141 is a subgroup of $A(G)$; we have

$$I(G) \subseteq A(G) \subseteq S(G).$$

7.3   Isomorphism

DEFINITION. Let G, H be groups. Then we say G and H are *isomorphic* and write $G \cong H$, if there exists an isomorphism $\theta: G \to H$.

Isomorphism is the analogue for groups of similarity for sets (section 3.8). Groups which are isomorphic are considered 'equivalent' from the point of view of group theory; the relation $\cong$ has the properties of an equivalence relation, i.e. (1) $G \cong G$ for any group G, (2) if $G \cong H$ then $H \cong G$, finally (3) $G \cong H$ and $H \cong K$ implies $G \cong K$. These facts are proved, using the lemmas (iv), (v) of section 7.2, just as the corresponding facts for similarity (p. 38).

Groups which are isomorphic are exact 'copies' of each other, as far as their binary operations are concerned. For example the groups $Z/3Z$ and $A(4)/V$ (Examples 136, 137, p. 84) evidently have the 'same' multiplication tables, although they have quite different elements. This 'sameness' comes from the fact that the map $\theta: Z/3Z \to A(4)/V$, which takes

$$E_0 \to V, \ E_1 \to A, \ E_2 \to P,$$

is an *isomorphism*, and this shows that $Z/3Z \cong A(4)/V$.

A property of groups is said to be *invariant* if whenever G, H are isomorphic groups and G has the property, then also H has it. The *order* of a finite group is invariant. For if $G \cong H$ then certainly $G \simeq H$ (since any isomorphism θ is bijective), and this means that isomorphic groups G, H have the same order. Group theory might be described as the study of properties of groups which are invariant in this sense.

Example 145. Example 138 shows that $(R, +) \cong (R^+, .)$.

Example 146. If $G \cong H$ and G is Abelian, then H is also Abelian. For let u, $v \in H$ and let $\theta: G \to H$ be an isomorphism. If $x = \theta^{-1}(u)$, $y = \theta^{-1}(v)$ then x, $y \in G$, hence $xy = yx$ since G is Abelian; now apply θ to this equation. By **H** we get $\theta(x)\theta(y) = \theta(y)\theta(x)$, i.e. $uv = vu$. Therefore H is Abelian; the property of being Abelian is invariant.

Example 147. The groups $S(3)$ and $Z/6Z$ both have order 6, but are not isomorphic, because $S(3)$ is not Abelian and $Z/6Z$ is.

Example 148. Let $\theta: G \to H$ be any isomorphism, and x, y elements of G, H respectively, such that $\theta(x) = y$ (hence also $\theta^{-1}(y) = x$). From Example 143 and section 7.2(ii) we find easily that $x^n = e_G \Leftrightarrow y^n = e_H$, for any integer n. Thus x, y have the same order. We can use this to show that the groups V (p 244) and $Z/4Z$ are not isomorphic, although both are of order 4 and both are Abelian. For $Z/4Z$ has an element $E_1 = 4Z + 1$ of order 4, and V has no element of order 4.

7.4 Kernel and image

Any homomorphism $\theta: G \to H$ gives some connection between the groups G and H. If θ is an isomorphism then this connection is very close, and we know that $G \cong H$. If θ is not an isomorphism we cannot say as much, but the fundamental 'Homomorphism Theorem'† below tells us that *any* homomorphism of G into H gives rise to an *isomorphism* of a certain quotient group of G and a certain subgroup of H.

DEFINITION. Let G, H be groups and $\theta: G \to H$ a homomorphism.

The sets

$$\text{Ker } \theta = \{x \in G \mid \theta(x) = e_H\}, \tag{7.1}$$

$$\text{Im } \theta = \{\theta(x) \mid x \in G\} \tag{7.2}$$

are called the *kernel* and *image* of θ, respectively.

Notice that Ker θ is a subset of G, while Im θ is a subset of H.

THE HOMOMORPHISM THEOREM. *Let G, H be groups and $\theta: G \to H$ any homomorphism. Then*

(i)   *Ker θ is a normal subgroup of G.*

(ii) *Im θ is a subgroup (not necessarily normal) of H.*

(iii) *If we write $K = \text{Ker } \theta$, there is an isomorphism*

$$\theta^*: G/K \to \text{Im } \theta$$

defined by the rule: if $x \in G$ then $\theta^*(Kx) = \theta(x)$. Hence

$$G/\text{Ker } \theta \cong \text{Im } \theta.$$

---

†Sometimes called the 'first isomorphism theorem'.

Proof. (i) Let x, $y \in \text{Ker } \theta$, so that $\theta(x) = \theta(y) = e_H$. Then by section 7.2(i) $\theta(xy^{-1}) = e_H e_H^{-1} = e_H$, hence $xy^{-1} \in \text{Ker } \theta$, and this shows that $\text{Ker } \theta$ is a subgroup of G. To prove it is normal in G, use condition **N''** (p. 81). If $u \in \text{Ker } \theta$ and x is *any* element of G, $\theta(x^{-1}ux) = \theta(x)^{-1}\theta(u)\theta(x) = \theta(x)^{-1}e_H\theta(x) = \theta(x)^{-1}\theta(x) = e_H$, so that $x^{-1}ux \in \text{Ker } \theta$, as required.

(ii) Let u, v be any elements of $\text{Im } \theta$, we want to prove that $uv^{-1} \in \text{Im } \theta$, i.e. that $\text{Im } \theta$ satisfies condition **S** (p. 58). By definition of $\text{Im } \theta$, there exist x, $y \in G$ such that $\theta(x) = u$ and $\theta(y) = v$. By section 7.2(i), $uv^{-1} = \theta(xy^{-1})$, which belongs to $\text{Im } \theta$.

(iii) Let x, y be any elements of G. We prove first a lemma.

LEMMA. $Kx = Ky \Leftrightarrow yx^{-1} \in K \Leftrightarrow \theta(x) = \theta(y)$.

The first implication is simply the theorem on p. 69, and for the second we have: $yx^{-1} \in K = \text{Ker } \theta \Leftrightarrow \theta(yx^{-1}) = e_H \Leftrightarrow \theta(y)\theta(x)^{-1} = e_H$ (by section 7.2(i))$\Leftrightarrow \theta(y) = \theta(x)$. This proves the lemma.

The *consistency* of the rule defining θ^* follows from the statement $yx^{-1} \in K \Rightarrow \theta(x) = \theta(y)$. From the statement $\theta(x) = \theta(y) \Rightarrow Kx = Ky$ we see θ^* *is injective.* From the definition of $\text{Im } \theta$, the map θ^* *is surjective.* We now show that θ^* satisfies **H**: let x, $y \in G$, then $\theta^*[(Kx)(Ky)] = \theta^*(Kxy) = \theta(xy) = \theta(x)\theta(y) = \theta^*(Kx)\theta^*(Ky)$. Therefore θ^* is a homomorphism. But we have just seen θ^* is bijective. Hence θ^* is an isomorphism.

The Homomorphism Theorem tells us that the kernel $\text{Ker } \theta$ of any homomorphism $\theta: G \to H$, is always a normal subgroup of G. We shall now prove that, conversely, *every normal subgroup N of a group G, is the kernel of some homomorphism from G to some group H*. Take H to be the quotient group G/N, and define a map

$$v: G \to G/N \tag{7.3}$$

by the rule: if $x \in G$, let $v(x) = Nx$. According to the definition of the group G/N (see section 6.8), we have

$$v(x)v(y) = (Nx)(Ny) = Nxy = v(xy),$$

hence v satisfies **H**; therefore v is a homomorphism. Now $\text{Ker } v$ is, by the definition of kernel, the set of all $x \in G$ such that $v(x) = Nx$ is the unit element of G/N, i.e. such that $Nx = N$. But $Nx = N \Leftrightarrow x \in N$ (see theorem, p. 69), hence Ker $v = N$, and this proves the statement printed in italics above.

Notice that v is surjective, i.e. it is an epimorphism. It is given a special name, by the following definition.

DEFINITION. If N is a normal subgroup of a group G, then the map (7.3) is called the *natural epimorphism* of G onto G/N.

Example 149. If $\theta: G \to H$ is injective, then the only element $x \in G$ such that $\theta(x) = e_H$ is $x = e_G$ (we know $\theta(e_G) = e_H$, by section 7.2(ii)), hence Ker $\theta = \{e_G\}$. Conversely, the lemma above shows that if Ker $\theta = \{e_G\}$, then $\theta(x) = \theta(y) \Rightarrow x = y$, i.e. θ is injective. Hence a *homomorphism θ is a monomorphism if and only if Ker $\theta = \{e_G\}$*. This is a companion to the theorem that a *homomorphism θ is an epimorphism* (i.e. *is surjective*) *if and only if Im $\theta = H$*, which is immediate from the definition of Im θ.

Example 150. If $\{e\}$ is the one-element subgroup of a group G, then $G/\{e\} \cong G$. This is very easy to verify directly, but it is also a consequence of the Homomorphism Theorem – take $\theta = \iota_G$, and check that Ker $\iota_G = \{e\}$, Im $\iota_G = G$.

Example 151. Cyclic groups. Let G be a cyclic group with unit element e, generated by an element x. Let $\theta: Z \to G$ be the homomorphism defined $\theta(n) = x^n (n \in Z)$, see Example 139. Then Im $\theta = \{x^n | n \in Z\} = \text{gp}\{x\} = G$, and Ker θ is the set all integers n such that $x^n = e$. If G infinite, then Ker $\theta = \{0\}$ (see p. 61); if G has finite order m, then (Example 105, p. 62) Ker θ is the group mZ of all multiples of m. The Homomorphism Theorem now shows: if G is infinite, then $Z/\{0\} \cong G$, hence $Z \cong G$; while if G has finite order m, we have $Z/mZ \cong G$.

Example 152. Alternating groups. Let $x_1, \ldots, x_n$ be n variables $(n \geq 2)$ and let $\Delta(x_1, \ldots, x_n)$ be the polynomial $\prod_{i < j}(x_i - x_j)$, e.g. $\Delta(x_1, x_2, x_3) = (x_1 - x_2)(x_1 - x_3)(x_2 - x_3)$. Then for each permutation θ of $\{1, 2, \ldots, n\}$, the polynomial $\theta\Delta = \Delta(x_{\theta(1)}, \ldots, x_{\theta(n)})$ is equal either to Δ or to $-\Delta$; θ is called *even* or *odd*, in these respective cases. Define a map $\varepsilon: S(n) \to R$ as follows: $\varepsilon(\theta) = 1$ if θ is even, $\varepsilon(\theta) = -1$ if θ is odd. Thus $\theta\Delta = \varepsilon(\theta)\Delta$ for all θ. Then ε *is a homomorphism of $S(n)$ into $(R^*, .)$*. For if θ, ϕ are any elements of $S(n)$, $(\theta\phi)\Delta = \theta[\varepsilon(\phi)\Delta] = \varepsilon(\phi)(\theta\Delta) = \varepsilon(\theta)\varepsilon(\phi)\Delta$, which shows that $\varepsilon(\theta\phi) = \varepsilon(\theta)\varepsilon(\phi)$, i.e. that ε satisfies **H**. Ker ε is the set $A(n)$ of all even permutations of $\{1, 2, \ldots, n\}$, and Im ε is the two-element set $\{1, -1\}$ (notice that the identity permutation is even, and $\theta = (12)$ is odd, see P. M. Cohn, *Linear Equations*). The Homomorphism Theorem now tells us that $A(n) \trianglelefteq S(n)$ and that $S(n)/A(n) \cong \{1, -1\}$, and in particular $|S(n)/A(n)| = 2$, i.e. $|A(n)| = \frac{1}{2}|S(n)| = \frac{1}{2}n!$, by Lagrange's theorem. $A(n)$ is called the *alternating group of degree n*. The multiplication table of $A(4)$ is given on p. 243.

Example 153. Cayley's Theorem. Let x be any element of a group G, and define the map $\lambda_x : G \to G$ by the rule: if $a \in G$ let $\lambda_x(a) = xa$. From the group axioms **G1, G2** follow at once (i) $\lambda_x \lambda_y = \lambda_{xy}$ for any $x, y \in G$, and (ii) $\lambda_e = \iota_G$, the identity map of G onto itself. These show that, for any $x \in G$, the map $\lambda_x : G \to G$ is bijective (since there is a map θ' such that $\lambda_x \theta' = \theta' \lambda_x = \iota_G$, namely $\theta' = \lambda_{x^{-1}}$; see section 3.6(iii), p. 34), i.e. λ_x is a permutation of the set G. But (i) also shows that the map $\lambda : G \to S(G)$ defined by the rule: if $x \in G$ let $\lambda(x) = \lambda_x$, is a homomorphism of G into the group $S(G)$ of all permutations of G onto itself. (Recall that $S(G)$ is the *symmetric group* on the set G, p. 51.) Ker λ consists of all $x \in G$ such that $\lambda_x = \iota_G$; but for such x, we have $xa = \iota_G(a)$ for all $a \in G$, hence $x = e$. Therefore Ker $\lambda = \{e\}$. Im λ is some subgroup, say G_0, of $S(G)$. The Homomorphism Theorem says $G/\{e\} \cong$ Im λ, i.e. $G \cong G_0$ (see Example 150). This proves Cayley's Theorem, that *any group is isomorphic to a subgroup of a symmetric group*.

7.5   Lattice diagrams

The rest of this chapter is concerned with connections between subgroups and homomorphisms. It will be useful to show the subgroups of a given group G on a diagram, called a *lattice diagram*, such as Figure 13. A lattice diagram is made up of 'points' (shown here as small circles), and 'lines' which join certain pairs of points. Each point represents a subgroup of G, and if the points representing subgroups X, Y are joined by a line, it means that either $X \subseteq Y$ or $X \supseteq Y$, according as the point X is lower or higher than the point Y. (The diagram never contains horizontal lines.) The diagram does not necessarily show *all* the subgroups of G; we are often interested in only a few subgroups, and to show more would make the diagram unnecessarily complicated. A diagram showing *all* the subgroups of G, is sometimes called the *(full) subgroup lattice* of G; for example Figure 13 shows all the

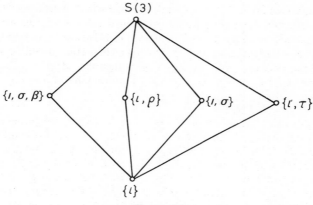

Figure 13

subgroups of $G = S(3)$. Notice that if G itself appears on the diagram, it must be at the *highest* point, since G contains all other subgroups of G; and if the 'unit subgroup' $\{e\}$ is on the diagram, it must be at the *lowest* point, since every other subgroup of G contains $\{e\}$. Figure 14 shows some of the subgroups of the group $Z = (Z, +)$ (see Example 154, below); in fact it shows all subgroups S of Z such that $S \supseteq 12Z$. Notice $mZ \subseteq nZ$ if and only if m is a multiple of n (see Example 155, below). For example the line joining $6Z$ and $2Z$ shows that $6Z \subseteq 2Z$.

Example 154. Subgroup of Z. We have often used the fact that, for any integer $m \geq 0$, the set $mZ = \{\ldots, -2m, -m, 0, m, 2m, \ldots\}$ is a subgroup of Z (see Example 96, p. 59). Now we shall prove the converse statement, namely that *if H is any subgroup of Z, then $H = mZ$ for some integer $m \geq 0$.* If $H = \{0\}$, then we can write $H = mZ$, with $m = 0$. If $H \neq \{0\}$, H contains some integer $r \neq 0$. Because H is a subgroup of Z, it also contains $-r$; therefore H contains some *positive* integer (since one of r or $-r$ must be positive!). Let m be the smallest positive integer contained in H. Then, because H is a subgroup, H must contain the subgroup mZ generated by m (see Example 103, p. 61). We shall now prove that every element of H lies in mZ; this will show $H \subseteq mZ$, and since also $H \supseteq mZ$, we shall have $H = mZ$, as we wanted to prove. Let x be any element of H. Divide x by m, so that $x = qm + r$, q being the quotient and r the remainder; note that $0 \leq r < m$. But $qm \in mZ$ and $mZ \subseteq H$, so that $qm \in H$. Therefore

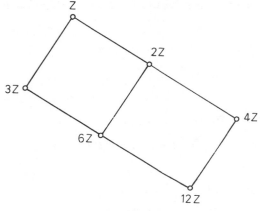

Figure 14

$x - qm = r \in H$ (because H is a subgroup and both $x, qm \in H$). If $r \neq 0$, it is a positive integer contained in H and smaller than m; this contradicts our definition of m as the smallest positive integer contained in H. So we must have $r = 0$. But this shows that $x = qm \in mZ$, which is what we needed to prove.

Example 155. Let m, n be positive integers. If m is an integral multiple of n, say $m = rn$ for some $r \in Z$, then $mZ \subseteq nZ$. For any element $x = zm$ of mZ ($z \in Z$) can be written $x = (zr)n$, which shows $x \in nZ$. Conversely, suppose $mZ \subseteq nZ$. Then $m \in mZ$, hence also $m \in nZ$, and so we must have $m = rn$ for some $r \in Z$. Thus $mZ \subseteq nZ$ if and only if m is a multiple of n, i.e. if and only if n divides m.

7.6 Homomorphisms and subgroups

In this section we assume given an epimorphism $\theta: F \to G$ between groups F and G; this means θ is a homomorphism, and Im $\theta = G$. Let $N = \text{Ker } \theta$; we know by the Homomorphism Theorem that $N \trianglelefteq G$ and $F/N \cong G$. We shall now show how the subgroups of G are related, by means of θ, to the subgroups of F which contain N.

THEOREM. *Let $\mathscr{S}$ be the set of all subgroups M of F such $M \supseteq N$, and* let $\mathscr{T}$ be the set of all subgroups H of G. Then there are maps

$$\Theta:\mathscr{S}\to\mathscr{T}, \qquad \Psi:\mathscr{T}\to\mathscr{S},$$

defined as follows:

If $M\in\mathscr{S}$, then $\Theta(M)=\{\theta(m)|m\in M\}$,

and

If $H\in\mathscr{T}$, then $\Psi(H)=\{f\in F|\theta(f)\in H\}$.

Moreover the maps Θ, Ψ are inverses of each other.

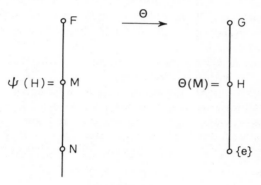

Figure 15

Proof. Let $M\in\mathscr{S}$. We want to prove that $\Theta(M)\in\mathscr{T}$, i.e. $\Theta(M)$ is a subgroup of G. Let $x_1, x_2\in\Theta(M)$. Then there exist $m_1, m_2\in M$ such that $\theta(m_1)=x_1$, $\theta(m_2)=x_2$. So $x_1 x_2^{-1}=\theta(m_1)\theta(m_2)^{-1}=\theta(m_1 m_2^{-1})$ (by section 7.2 lemma (i)), and $m_1 m_2^{-1}\in M$ because M is a subgroup of F (see **S**, p. 58). Hence $x_1 x_2^{-1}=\theta(m_1 m_2^{-1})$ belongs to $\Theta(M)$. This proves that $\Theta(M)$ is a subgroup of G.

Conversely if $H\in\mathscr{T}$, we must prove that $\Psi(H)\in\mathscr{S}$, i.e. that $\Psi(H)$ is a subgroup of F, and that $\Psi(H)\supseteq N$. Suppose $f_1, f_2\in\Psi(H)$. Then $\theta(f_1), \theta(f_2)\in H$, hence $\theta(f_1 f_2^{-1})=\theta(f_1)\theta(f_2)^{-1}\in H$ because H is a subgroup; therefore $f_1 f_2^{-1}\in\Psi(H)$, and this proves that $\Psi(H)$ is a subgroup of F. Now let n be any element of $N=\operatorname{Ker}\theta$. Then $\theta(n)=e_G$, which belongs to H, therefore $n\in\Psi(H)$. So $N\subseteq\Psi(H)$, and this proves that $\Psi(H)\in\mathscr{S}$.

We have now established that the maps $\Theta: \mathscr{S} \to \mathscr{T}$ and $\Psi: \mathscr{T} \to \mathscr{S}$ both exist. Next we shall prove that

$$\Theta[\Psi(H)] = H, \qquad \text{for all } H \in \mathscr{T}, \tag{7.4}$$

and

$$\Psi[\Theta(M)] = M, \qquad \text{for all } M \in \mathscr{S}. \tag{7.5}$$

First let $H \in \mathscr{T}$, and let h be any element of H. Because θ is surjective, there is some $f \in F$ such that $\theta(f) = h$. But this equation shows that $f \in \Psi(H)$. Hence $h = \theta(f) \in \Theta[\Psi(H)]$, and we have proved that $H \subseteq \Theta[\Psi(H)]$. For the converse inclusion, start with any element $h \in \Theta[\Psi(H)]$. Then $h = \theta(f)$, for some $f \in \Psi(H)$. But, because $f \in \Psi(H)$, we see $h = \theta(f) \in H$. Thus $\Theta[\Psi(H)] \subseteq H$, and therefore (7.4) is proved.

To prove (7.5) we can use a trick. Suppose $M \in \mathscr{S}$, so that $\Theta(M) \in \mathscr{T}$. Putting $H = \Theta(M)$ in (7.4), we get $\Theta(\Psi[\Theta(M)]) = \Theta(M)$. But we shall see from Example 156, below, that Θ is an injective map. Therefore $\Psi[\Theta(M)] = M$, and (7.5) is proved. This also completes the proof of the theorem, for (7.4) and (7.5) show that $\Theta\Psi =$ the identity map on $\mathscr{T}$, and $\Psi\Theta =$ the identity map on $\mathscr{S}$. Therefore Θ, Ψ are both bijective, and are inverse to each other (see section 3.6(iii), p. 34).

Example 156. Suppose M_1, $M_2 \in \mathscr{S}$, so that $\Theta(M_1)$, $\Theta(M_2) \in \mathscr{T}$. Then we have a useful result, namely

$$M_1 \subseteq M_2 \Leftrightarrow \Theta(M_1) \subseteq \Theta(M_2). \tag{7.6}$$

First assume $M_1 \subseteq M_2$. Any element $h \in \Theta(M_1)$ has the form $h = \theta(m_1)$, for some $m_1 \in M_1$. Since $M_1 \subseteq M_2$, $m_1 \in M_2$. Hence $h = \theta(m_1) \in \Theta(M_2)$. So $M_1 \subseteq M_2 \Rightarrow \Theta(M_1) \subseteq \Theta(M_2)$, which is one half of (7.6). To prove the other half, assume $\Theta(M_1) \subseteq \Theta(M_2)$; this time we must *prove* that $M_1 \subseteq M_2$. If $m_1 \in M_1$, then $\theta(m_1) \in \Theta(M_1)$, hence by our assumption $\theta(m_1) \in \Theta(M_2)$. So there is some $m_2 \in M_2$ such that $\theta(m_1) = \theta(m_2)$. Then $\theta(m_1 m_2^{-1}) = \theta(m_1)\theta(m_2)^{-1} = e_G$, i.e. $m_1 m_2^{-1} \in N = \text{Ker } \theta$. However, $N \subseteq M_2$ because $M_2 \in \mathscr{S}$. Thus $m_1 = (m_1 m_2^{-1})m_2 \in M_2$ because both $m_1 m_2^{-1}$ and m_2 belong to M_2. We have now shown that $m_1 \in M_1 \Rightarrow m_1 \in M_2$, i.e. that $M_1 \subseteq M_2$, as required. This completes the proof of (7.6).

One consequence of (7.6) is a fact we have used in the proof of the theorem above, namely that Θ is an injective map. For if $\Theta(M_1) = \Theta(M_2)$, we have $\Theta(M_1) \subseteq \Theta(M_2)$ *and* $\Theta(M_2) \subseteq \Theta(M_1)$. So (7.6) gives us $M_1 \subseteq M_2$ and $M_2 \subseteq M_1$, i.e. $M_1 = M_2$.

Expression (7.6) has a useful 'pictorial' interpretation, namely it shows that Θ 'maps' the lattice diagram of $\mathscr{S}$, to that of $\mathscr{T}$; for example, if the points representing M_1, M_2 are joined in the diagram of $\mathscr{S}$, it means that either $M_1 \subseteq M_2$ or $M_2 \subseteq M_1$. *Then* (7.6) *shows that either* $\Theta(M_1) \subseteq \Theta(M_2)$ *or* $\Theta(M_2) \subseteq \Theta(M_1)$, so the points representing $\Theta(M_1)$, $\Theta(M_2)$ are joined in the diagram of $\mathscr{T}$; similarly any line in the diagram of $\mathscr{T}$, corresponds to a line in the diagram of $\mathscr{S}$. This explains why the lattice diagrams in Figures 14 and 16 are identical, apart from the labelling of the points; Figure 14 is the diagram for $\mathscr{S}$ (see Example 157 for an explanation of the notation) and Figure 16 is the (full) subgroup lattice of G, i.e. it is the diagram of $\mathscr{T}$. The map Θ takes each subgroup shown in Figure 14, to the subgroup occupying the corresponding positions in Figure 16, e.g. $\Theta(6Z) = gp\{x^6\}$, $\Theta(Z) = G$, etc.

Example 157. Subgroups of a finite cyclic group. Let G be a cyclic group with finite order m, generated by an element x. Let $\theta: Z \to G$ be the

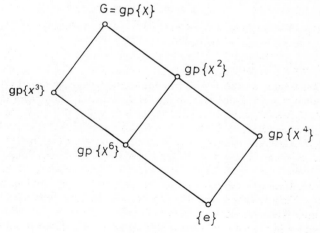

Figure 16

homomorphism given by $\theta(n) = x^n (n \in Z)$. In Example 151 we found that Im $\theta = G$ (i.e. θ is an epimorphism) and Ker $\theta = mZ$. So the theorem above tells us there is a bijective map $\Theta : \mathscr{S} \to \mathscr{T}$, where $\mathscr{T}$ is the set of all subgroups of G, and $\mathscr{S}$ is the set of all subgroups of Z which contain mZ. By Examples 154, 155, $\mathscr{S}$ consists of all subgroups $nZ(n \geq 0)$ such that n divides m. By the definition of Θ, $\Theta(nZ) = \{\theta(f) | f \in nZ\} = \{x^f | f = ns$ for some $s \in Z\} = \{(x^n)^s | s \in Z\} = \text{gp}\{x^n\}$. So $\mathscr{T}$, the set of all subgroups of G, consists of the subgroups $\text{gp}\{x^n\}$, for all integers $n(\geq 0)$ which divide m. Notice that $\text{gp}\{x^n\}$ is cyclic, and has order $d = m/n$, since the elements of $\text{gp}\{x^n\}$ are $e, x^n, (x^n)^2, \ldots, (x^n)^{d-1}$. So we can restate our conclusion: G has exactly one subgroup of each order d which divides m; this subgroup is cyclic, and is $\text{gp}\{x^n\}$ where $n = m/d$.

7.7 The second isomorphism theorem

Suppose that G is a group, written multiplicatively, and with unit element e. If U, V are any subsets of G, define UV to be the subset $UV = \{uv | u \in U, v \in V\}$. For example if $G = S(3)$, we might take $U = \{\alpha, \beta\}$, $V = \{\iota, \beta, \rho\}$ in the notation of p. 241. Then $UV = \{\alpha\iota, \alpha\beta, \alpha\rho, \beta\iota, \beta\beta, \beta\rho\} = \{\alpha, \iota, \tau, \beta, \alpha, \sigma\} = \{\iota, \alpha, \beta, \sigma, \tau\}$.

We have already used a slight variant of this notation, in the case where U or V is a single element set. For example if $V = \{x\}$, then $U\{x\}$ is the same as the set Ux as defined in section 6.1, p. 68. This 'multiplication' of subsets of G is associative, i.e. $(UV)W = U(VW)$ for any subsets U, V, W of G. For $(UV)W = \{(uv)w | u \in U, v \in V, w \in W\} = \{u(vw) | u \in U, v \in V, w \in W\} = U(VW)$.

Now suppose that U, V are both *subgroups* of G. It is important to realize that UV is not necessarily a subgroup. For example, the sets $U = \{\iota, \sigma\}$ and $V = \{\iota, \tau\}$ are both subgroups of $G = S(3)$, but $UV = \{\iota, \sigma, \tau, \alpha\}$ is not a subgroup. However there is an important case where the product of subgroups U, V of G *is* a subgroup.

LEMMA. *If U, V are subgroups of G, and if either U or V is normal in G, then UV is a subgroup of G.*

Proof. We shall prove that UV is a subgroup of G, in the case that $U \trianglelefteq G$. (The proof for the case $V \trianglelefteq G$ is very similar, and is left to the reader.) Let x_1, x_2 be any elements of UV. Then $x_1 = u_1 v_1$, $x_2 = u_2 v_2$, for certain u_1, $u_2 \in U$, v_1, $v_2 \in V$. Hence $x_1 x_2^{-1} =$

$u_1 v_1 v_2^{-1} u_2^{-1}$. Because V, U are subgroups, the element $v = v_1 v_2^{-1} \in V$, and $u_2^{-1} \in U$. Hence $v u_2^{-1} v^{-1} \in U$ because $U \trianglelefteq G$ (see **N″**, p. 81).   Thus $x_1 x_2^{-1} = u_1 v u_2^{-1} = u_1 (v u_2^{-1} v^{-1}).v \in UV$,   since $u_1 (v u_2^{-1} v^{-1}) \in U$ and $v \in V$. This shows that UV is a subgroup of G.

COROLLARY. *If U, V are any subgroups of an Abelian group, then UV is a subgroup. For any subgroup of an Abelian group G, is normal in G.*

From now on we make a slight change of notation, and take N, H to be subgroups of G, with $N \trianglelefteq G$. The lemma shows that NH is a subgroup, It is clear that $NH \supseteq N$ and $NH \supseteq H$; moreover if S is any subgroup of G which contains both N and H, then $S \supseteq N$ (for S contains every element $n \in N$ and every $h \in H$, hence S contains every element $nh \in NH$). It is interesting to compare NH with the subgroup† $N \cap H$, which is contained in both N and H; moreover if T is any subgroup contained in both N and H, then $T \subseteq N \cap H$. The lattice diagram of the four subgroups N, H, NH, $N \cap H$ is shown in Figure 17. The following theorem is sometimes called the 'second isomorphism theorem'.

THEOREM. *Let N, H be subgroups of a group G, with N normal in G.*

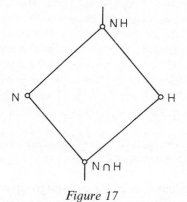

Figure 17

---

†Notice that $N \cup H$ is usually not a subgroup. See Chapter 5, Exercise 7.

Then $N \trianglelefteq NH$, $N \cap H \trianglelefteq H$ and

$$H/N \cap H \cong NH/N.$$

Proof. Since $N \trianglelefteq G$ we have $x^{-1}Nx = N$ for all $x \in G$ (see **N'**, p. 81). But then $x^{-1}Nx = N$ for all $x \in N$, since $NH \subseteq G$. So $N \trianglelefteq NH$.

The elements of the quotient group NH/N are the cosets Nx, for $x \in NH$. Each $x \in NH$ can be written $x = nh$, with $n \in N$, $h \in H$. Since $Nn = N$ for any $n \in N$, we have $Nx = Nnh = Nh$. Thus

$$NH/N = \{Nh | h \in H\}. \tag{7.7}$$

Let $v: G \to G/N$ be the 'natural epimorphism' (p. 93), so that $v(x) = Nx$, for all $x \in G$. Define $\theta: H \to G/N$ to be the 'restriction of v to H' i.e. θ is the map with domain H, which has the same effect on elements of H, as v does. In other words, $\theta(h) = Nh$, for all $h \in H$. Clearly θ is a homomorphism, because v is. Also Ker $\theta = H \cap N$ (since $N = $ Ker v). And by (7.7) we see that Im $\theta = NH/N$. So the Homomorphism Theorem, applied to θ, tells us that $H \cap N \trianglelefteq H$ and that $H/N \cap N \cong NH/N$. This proves the theorem.

Example 158. Figure 17 helps to remember this theorem. The quotient group NH/N can be represented by the line joining the point NH to the point N, similarly $H/H \cap N$ by the line joining H to $H \cap N$. The theorem then says that these opposite sides of the parallelogram are isomorphic. But since H may not be normal in NH, we do not in general get an isomorphism $NH/N \cong N/H \cap N$ from the other pair of opposite sides.

Example 159. Suppose U, V are subsets of an Abelian group G which is written in *additive* notation. We define $U + V = \{u + v | u \in U, v \in V\}$; this is a subset of G. The lemma above (now expressed in additive notation) shows that *if U, V are subgroups of G, then $U + V$ is also a subgroup of G* – remember that *every* subgroup of an Abelian group G is normal in G. We may take $N = U$ in the theorem, and get $U + V/U \cong V/U \cap V$. In this case we have also $U + V/V \cong U/U \cap V$, since V is normal in G.

7.8 Direct products and direct sums of groups

Let G, H be groups, both written multiplicatively, and with unit elements e_G, e_H, respectively. The product set $G \times H$ (see section 1.7) has for its elements all pairs (g, h), with $g \in G, h \in H$. We can define a binary operation, or multiplication, on $G \times H$ by the rule

$$(g_1, h_1)(g_2, h_2) = (g_1 g_2, h_1 h_2), \tag{7.8}$$

for all $g_1, g_2 \in G$ and $h_1, h_2 \in H$. With this operation, $G \times H$ becomes a group, which is called the *direct product of the groups G and H*.

We must check that the group axioms are satisfied. Let $x = (g_1, h_1)$, $y = (g_2, h_2)$, $z = (g_3, h_3)$ be any elements of $G \times H$.

G1. $(xy)z = [(g_1 g_2)g_3, (h_1 h_2)h_3]$, and $x(yz) = [g_1(g_2 g_3), h_1(h_2 h_3)]$, using the multiplication rule (7.8). But these elements are equal, since $(g_1 g_2)g_3 = g_1(g_2 g_3)$ and $(h_1 h_2)h_3 = h_1(h_2 h_3)$, because the associative law holds in both G and H.

G2. The element $e = (e_G, e_H)$ is a unit element for $G \times H$, since $xe = (g_1 e_G, h_1 e_H) = (g_1, h_1) = x$, and similarly $ex = x$, for all $x = (g_1, h_1) \in G \times H$.

G3. The element $x = (g_1, h_1)$ has inverse $x^{-1} = (g_1^{-1}, h_1^{-1})$. For $xx^{-1} = (g_1 g_1^{-1}, h_1 h_1^{-1}) = (e_G, e_H) = e$, and similarly $x^{-1}x = e$. Thus the set $G \times H$, with the binary operation given by (7.8), is indeed a group.

If G, H are both finite, then so is $G \times H$, and its order is $|G \times H| = |G| \, |H|$ (see Example 22, p. 10). If G, H are both Abelian, then so is $G \times H$; this follows at once from (7.8).

Example 160. There is nothing to stop us from taking $G = H$; of course in that case the direct product $G \times G$ has as elements all pairs (g_1, g_2) with $g_1, g_2 \in G$. For example let C_2 denote cyclic group of order 2, generated by an element x. Then $x^2 = e$, and $C_2 = \{e, x\}$. The direct product $C_2 \times C_2$ has four elements, *viz.* (e, e), (x, e), (e, x), (x, x). The reader can check the multiplication table below, using the multiplication rule (7.8).

(e, e)	(x, e)	(e, x)	(x, x)
(x, e)	(e, e)	(x, x)	(e, x)
(e, x)	(x, x)	(e, e)	(x, e)
(x, x)	(e, x)	(x, e)	(e, e)

From this table we can see that $C_2 \times C_2 \cong V$. In fact the bijective map $\theta: C_2 \times C_2 \to V$ which takes $(e, e) \to 1$, $(x, e) \to t$, $(e, x) \to u$, $(x, x) \to v$ is an isomorphism, as one sees by comparing the table above with the multiplication table for V (p. 244).

Example 161. Direct products can be defined for any number of groups. If $G_1, \ldots, G_n$ are groups, define $G = G_1 \times \ldots \times G_n$ to be the set of all 'n-tuples' or 'n-vectors' (see p. 11) $(g_1, \ldots, g_n)$ with $g_i \in G_i (i = 1, \ldots, n)$, with multiplication

$$(g_1, \ldots, g_n)(h_1, \ldots, h_n) = (g_1 h_1, \ldots, g_n h_n),$$

for all g_i, $h_i \in G_i$ $(i = 1, \ldots, n)$. Verification of the group axioms is straightforward. G has unit element $e = (e_1, \ldots, e_n)$, where e_i is the unit element of $G_i (i = 1, \ldots, n)$. The inverse of $g = (g_1, \ldots, g_n)$ is $g^{-1} = (g_1^{-1}, \ldots, g_n^{-1})$.

Taking the case $n = 3$, notice that we have now defined three groups

$$(G_1 \times G_2) \times G_3, \; G_1 \times (G_2 \times G_3), \; G_1 \times G_2 \times G_3, \tag{7.9}$$

which are all different. However, they are isomorphic. For example define $\theta: (G_1 \times G_2) \times G_3 \to G_1 \times G_2 \times G_3$ to be the map which sends $[(g_1, g_2), g_3] \to (g_1, g_2, g_3)$ for all $g_i \in G_i$ $(i = 1, 2, 3)$; it is easy to check that θ is an isomorphism. There is a similarly 'natural' isomorphism from $G_1 \times (G_2 \times G_3)$ to $G_1 \times G_2 \times G_3$. For this reason it is customary – although strictly incorrect! – to regard the three groups (7.9) as identical, i.e. to 'omit brackets' in such groups as $(G_1 \times G_2) \times G_3$.

Let G, H be groups with unit elements e_G, e_H, respectively. Then G, H are *not* subgroups of the direct product $G \times H$, but we shall now show that $G \times H$ contains certain subgroups which are 'isomorphic copies' of G and H. Define

$$\tilde{G} = G \times \{e_H\} = \{(g, e_H) | g \in G\},$$

and

$$\tilde{H} = \{e_G\} \times H = \{(e_G, h) | h \in H\}.$$

THEOREM 1.

(i) $\tilde{G}, \tilde{H}$ are both normal subgroups of $G \times H$.

(ii) $G \cong \tilde{G}, H \cong \tilde{H}$.

(iii) $\tilde{G} \cap \tilde{H} = \{e\}$, where $e = (e_G, e_H)$ is the unit element of $G \times H$.

(iv) $\tilde{G}\tilde{H} = G \times H$.

Proof. (i) Let $x_1 = (g_1, e_H), x_2 = (g_2, e_H)$ be any elements of $\tilde{G}$. Then $x_1 x_2^{-1} = (g_1 g_2^{-1}, e_H e_H^{-1}) = (g_1 g_2^{-1}, e_H)$ belongs to $\tilde{G}$; hence $\tilde{G}$ is a subgroup. For any $z = (g, h)$ in $G \times H$ we have $z^{-1} x_1 z = (g^{-1} g_1 g, h^{-1} e_H h) = (g^{-1} g_1 g, e_H) \in \tilde{G}$. Hence $\tilde{G}$ is a normal subgroup of $G \times H$ (see **N''**, p. 81). One proves similarly that $\tilde{H} \trianglelefteq G \times H$.

(ii) Define the map $\theta: G \to \tilde{G}$ by the rule $\theta(g) = (g, e_H)$, for all $g \in G$. It is easy to check that θ is bijective. Also $\theta(g_1 g_2) = (g_1 g_2, e_H) = (g_1, e_H)(g_2, e_H) = \theta(g_1)\theta(g_2)$, which shows that θ is an isomorphism; hence $G \cong \tilde{G}$. One proves similarly that $H \cong \tilde{H}$.

(iii) If $(g, h) \in \tilde{G} \cap \tilde{H}$, we have $h = e_H$ because $(g, h) \in \tilde{G}$, and also $g = e_G$ because $(g, h) \in \tilde{H}$. Hence $(g, h) = e$.

(iv) $\tilde{G}\tilde{H}$ is a subset of $G \times H$. But each element (g, h) of $G \times H$ can be written $(g, h) = (g, e_H)(e_G, h)$, showing that $(g, h) \in \tilde{G}\tilde{H}$. Therefore $\tilde{G}\tilde{H} = G \times H$.

This theorem has the following converse, which is often very useful, for it shows how to recognize that a given group P is isomorphic to a direct product.

THEOREM 2. *Let P be a group, written multiplicatively, and with unit element e. Suppose that P has subgroups U, V which satisfy the following three conditions:*

(a) *U, V are both normal in P.*

(b) *$U \cap V = \{e\}$, and*

(c) *$UV = P$.*

Then P is isomorphic to the direct product $U \times V$.

Proof. Define a map $\theta: U \times V \to P$ by the rule, $\theta(u, v) = uv$, for all $u \in U, v \in V$. We shall prove that θ is an isomorphism. First prove that θ *is bijective*. Condition (c) shows that θ is surjective. Condition (b) shows that θ is injective, for if (u_1, v_1), (u_2, v_2) are elements of $U \times V$ such that $\theta(u_1, v_1) = \theta(u_2, v_2)$, we have $u_1 v_1 = u_2 v_2 \Rightarrow u_2^{-1} u_1 = v_2 v_1^{-1}$. But this shows that the element $p = u_2^{-1} u_1 = v_2 v_1^{-1}$ lies in $U \cap V$, so $p = e$, by (b); hence $u_1 = u_2$ and $v_1 = v_2$, i.e. $(u_1, v_1) = (u_2, v_2)$. It remains to prove that θ *is a homomorphism*. So far, we have not used condition (a). But we shall see in Example 162, below, that conditions (a) and (b) together imply that every element of U commutes with every element of V. So if $x = (u_1, v_1)$, $y = (u_2, v_2)$ are any elements of $U \times V$, then $xy = (u_1 u_2, v_1 v_2)$ and hence $\theta(xy) = u_1 u_2 v_1 v_2 = u_1 v_1 u_2 v_2 = \theta(x)\theta(y)$. This proves that θ is a homomorphism. Thus θ is an isomorphism, which completes the proof of the theorem.

DEFINITION. If P is a group having subgroups U, V which satisfy conditions (a), (b), (c) above, P is said to be the *internal direct product* of U and V. The theorem just proved, says that in this case, P is isomorphic to (but is not the same as) the direct product $U \times V$. (For this reason, the direct product $U \times V$ is sometimes called the *external* direct product of U and V.)

Example 162. Commutators. Let P be a multiplicatively written group with unit element e, and let u, v be any elements of P. The element $u^{-1} v^{-1} uv$ is called the *commutator of u and v*, and is often denoted $[u, v]$. It is easy to check that $[u, v] = e \Leftrightarrow uv = vu$.

Now suppose that U, V are normal subgroups of P, such that $U \cap V = \{e\}$. Take any elements $u \in U, v \in V$. Since $U \trianglelefteq P, v^{-1} uv \in U$, hence $[u, v] = u^{-1}(v^{-1} uv) \in U$. But since $V \trianglelefteq P$, $u^{-1} v^{-1} u \in V$, hence $[u, v] = (u^{-1} v^{-1} u) v \in V$. Therefore $[u, v] \in U \cap V = \{e\}$, so that $[u, v] = e$, i.e. u and v commute. This fact was used in the proof above.

Direct sums

If G, H are written with *additive* notation, with unit elements 0_G, 0_H, it is customary to use additive notation also for the direct product $G \times H$. The rule (7.8) now reads

$$(g_1, h_1) + (g_2, h_2) = (g_1 + g_2, h_1 + h_2),$$

for all $g_1, g_2 \in G$, $h_1, h_2 \in H$. The unit element of $G \times H$ is now written as $0 = (0_G, 0_H)$; the 'inverse' of (g, h) is $-(g, h) = (-g, -h)$. In these circumstances $G \times H$ is sometimes called the *direct sum of G and H*, and denoted $G \dotplus H$ or $G \oplus H$. It should be emphasized that these are merely alternative descriptions of the direct product; the elements are still pairs (g, h), with $g \in G$, $h \in H$. We shall not use the notations $G \dotplus H$, $G \oplus H$ for the direct product.

Example 163. Let $G_1, \ldots, G_n$ be groups, additively written. Then the direct product $G = G_1 \times \ldots \times G_n$ (see Example 161) may be written in additive notation; the rule of addition is

$$(g_1, \ldots, g_n) + (h_1, \ldots, h_n) = (g_1 + h_1, \ldots, g_n + h_n),$$

for all $g_i, h_i \in G_i$ $(i = 1, \ldots, n)$. A special case of great importance occurs when $G_1 = G_2 = \ldots = G_n = R$, the additive group of real numbers. The direct product $R \times \ldots \times R$ is usually written as R^n. Its elements are the ordinary 'n-vectors', and the rule just given, is the usual rule for vector addition. We shall return to this group R^n in Chapter 9.

Internal direct sums

It is worthwhile re-writing Theorem 2 in additive notation. Remember that any subgroup of an Abelian group is normal so we do not need the condition (a) of Theorem 2.

THEOREM 2′. *Let P be an Abelian group, written additively, and with unit element 0. Suppose that P has subgroups U, V which satisfy the following two conditions:*
(a) $U \cap V = \{0\}$, *and*
(b) $U + V = P$.
Then P is isomorphic to the direct product (or sum) $U \times V$.

Of course, we do not need a proof of this; it is a special case of Theorem 2. The proof of Theorem 2, expressed 'additively', shows that there is an isomorphism $\theta: U \times V \rightarrow P$, given by the rule $\theta(u, v) = u + v$, for all $u \in U$, $v \in V$.

DEFINITION. If P is an additive Abelian group having subgroups U, V which satisfy conditions (a), (b) above, P is said to be the *internal direct sum* of U and V, and we write $P = U \oplus V$ in this case. By what has just been said, P is isomorphic to (but is not the same as) the direct product (or sum) $U \times V$. Notice that if $P = U \oplus V$, then each element $p \in P$ has a *unique* expression $p = u + u$, with $u \in U$, $v \in V$.

Exercises for Chapter 7

1 If G, H are any groups, define the map $\theta: G \to H$ by the rule $\theta(x) = e_H$, for all $x \in G$. Prove that θ is a homomorphism.

2 If G is Abelian and n any integer, define $\theta: G \to G$ by the rule $\theta(x) = x^n$, for all $x \in G$. Prove that θ is an endomorphism of G.

3 Prove that $Z \cong mZ$, for any integer $m \neq 0$.

4 Prove that $(R^*, .)$ and $(R, +)$ are *not* isomorphic. (Hint: notice that $(R^*, .)$ has an element of order 2, *viz.* $x = -1$. If there were an isomorphism $\theta: R^* \to R$, what could $\theta(x)$ be?)

5 Let G be any group of order 4. Prove that either (i) G is cyclic (in which case $G \cong Z/4Z$ by Example 151) or (ii) $G \cong V$ (see p. 244). (*Hint*: the order of any element x of G must divide 4, by Example 122, p. 73. If G has an element x of order 4, then $G = \text{gp}\{x\}$ is cyclic. So if G is not cyclic, the order of each element $x \neq e$ must be 2. Try to make the multiplication table for G in this case.)

6 For any group G, define the map $\theta: G \to A(G)$ by the rule $\theta(x) = \pi_{x^{-1}}$ (see Examples 141 and 144 for notation), for all $x \in G$. Prove that θ is a homomorphism, and that Ker $\theta = Z(G)$ (the centre of G, p. 78), and Im $\theta = I(G)$ (see Example 141). Show that $G/Z(G) \cong I(G)$.

7 Let C be the set of all complex numbers, and $C^* = C - \{0\}$. Show that C^* is a multiplicative group. Now let R denote the *additive* group of all real numbers. Show that the map $\theta: R \to C^*$ defined by $\theta(r) = e^{2\pi i r}$ for all $r \in R$, is a homomorphism. Find Im θ and Ker θ, and deduce that $R/Z \cong U$, where U is the 'unit circle', i.e. the set of all complex numbers z which satisfy $|z| = 1$. (Note: this exercise requires some knowledge of complex numbers. See for example W. Ledermann, *Complex Numbers*, Library of Mathematics, Routledge & Kegan Paul.)

8 Draw the full subgroup lattice for cyclic groups of orders (i) 16, (ii) 18, (iii) 36. (Hint: use the method of Examples 156, 157, which shows that the full subgroup lattice of a cyclic group of order m, can be found from the lattice diagram of those subgroups of Z which contain mZ.)

9 Draw the full subgroup lattice of $D(8)$ (p. 244).

10 With the notation of section 7.6, prove that $\theta(M) \cong M/N$, for any $M \in \mathscr{S}$.

11 With the notation of section 7.6, let $M \in \mathscr{S}$ and $H = \Theta(M)$. Prove that M is a normal subgroup of F, if and only if H is a normal subgroup of G. Prove that in this case, $F/M \cong G/H$.

12 Let U, V be subgroups of a (multiplicatively written) group G. Prove that if UV is a subgroup of G, then $UV = VU$.

13 Prove the converse to the above, i.e. that if U, V are subgroups of G such that $UV = VU$, then UV is a subgroup of G.

14 Show that the subgroup $V = \{1, t, u, v\}$ (p. 244) is normal in $S(4)$ (use the method of Example 130 to calculate the elements $\pi \alpha \pi^{-1}$, for $\pi \in S(4)$ and $\alpha \in V$). Let H be the set of all $\pi \in S(4)$ such that $\pi(4) = 4$. Prove that H is a subgroup of $S(4)$, and also that $H \cong S(3)$. Use the theorem on p. 101 to prove that $S(4)/V \cong S(3)$.

15 If G, H are any groups, prove that $G \times H \cong H \times G$.

16 If G, H are any groups, define maps $\pi: G \times H \to G$ and $\theta: G \to G \times H$ by $\pi(g, h) = g$ and $\theta(g) = (g, e_H)$, for $g \in G$, $h \in H$. Prove that π and θ are both homomorphisms, and that Ker π is the subgroup $\tilde{H}$ of section 7.8, Theorem 1. Prove $(G \times H)/\tilde{H} \cong G$. Prove also that $\pi\theta = \iota_G$.

17 Let C_n denote a cyclic group of order n, for any integer $n \geq 1$. Prove that if m, n are integers with highest common factor 1, then $C_m \times C_n \cong C_{mn}$.

18 Let $R^2 = R \times R$, written additively. Then the elements (x, y) of R^2, represent points in the plane, by the usual convention of co-ordinate geometry. Show that any line L through the origin is a subgroup of R^2, and that the cosets of L in R^2 are the lines parallel to L.

19 With the notation of Exercise 18, suppose that L, M are two lines through the origin, and that $L \cap M = \{0\}$. Prove that $R^2 = L \oplus M$.

8 Rings and fields

8.1 Definition of a ring. Examples

In Chapter 4, section 4.4, we met certain *algebraic structures*, namely gruppoids, semigroups and groups. Each of these is a pair $(G, \circ)$, where G is a set and $\circ$ is a binary operation on G; the difference between a gruppoid, a semigroup and a group lies in the different conditions imposed on this operation $\circ$. We shall now study a new kind of algebraic structure which has *two* binary operations, which correspond to the two elementary operations of arithmetic, addition and multiplication.

DEFINITION. A *ring* is a triple $(S, +, \cdot)$, where S is a set, and

$$+ : S \times S \to S, \qquad \cdot : S \times S \to S$$

are two binary operations on S, known respectively as the *addition* and the *multiplication* in S. These operations map a pair $(x, y) \in S \times S$ to elements of S denoted respectively by $x + y$, xy. They must satisfy the following eight *ring axioms* **R1–R8**. To make these easier to remember, they are collected into three 'blocks' I, II and III.

I. These axioms involve only the addition operation, and they say that $(S, +)$ is an Abelian (i.e. commutative) group.

R1 $(x + y) + z = x + (y + z)$,   for all $x, y, z \in S$.

R2   There is an element 0 of S, such that $x + 0 = x = 0 + x$, for all $x \in S$.

R3   To each $x \in S$ there is an element $-x \in S$, such that $x + (-x) = 0 = (-x) + x$.

R4 $x + y = y + x$,   for all $x, y \in S$.

II. These axioms involve only the multiplication operation, and they say that $(S, \cdot)$ is a semigroup with unit element.

R5 $(xy)z = x(yz)$, for all $x, y, z \in S$.

R6 There is an element 1 of S, such that $x1 = x = x1$, for all $x \in S$.

III. These axioms connect the addition and multiplication operations. They are known as the *distributive laws*.

R7 $(x + y)z = xz + yz$,

and

R8 $x(y + z) = xy + xz$, for all $x, y, z \in S$.

From the discussion of associative operations in section 4.2, the 'general associative law' holds for both addition and multiplication; this means that expressions such as $x_1 + x_2 + \cdots + x_n$ and $x_1 x_2 \ldots x_n$ are independent of the way they are bracketed. In the case where $x_1, x_2, \ldots, x_n$ are all equal to an element x of S, then $x_1 + x_2 + \cdots + x_n$ is usually written nx, and $x_1 x_2 \ldots x_n$ is usually written x^n (see section 4.2). The usual conventions for an additively written Abelian group apply the ring S, e.g. we write $x - y$ for $x + (-y)$, and $(-n)x = n(-x)$ for a positive integer n and any $x \in S$ (see section 4.5).

From the discussion of units in section 4.3, we know that the elements 0, 1 in a ring are *unique*, i.e. 0 is the only element of S which satisfies axiom **R2**, and 1 is the only element of S which satisfies axiom **R6**. In the next section we shall prove that 0 is always a 'zero' for multiplication, i.e. that $x0 = 0 = 0x$ for all $x \in S$ (this follows from the ring axioms **R1–R8**; it does not have to be assumed as an extra axiom).

Notice that the addition in a ring is assumed to be commutative (axiom **R4**), but multiplication may be non-commutative.

DEFINITION. A ring $(S, +, \cdot)$ is said to be *commutative* if the multiplication in S is commutative, i.e. if S satisfies, in addition to the usual ring axioms **R1–R8**, the extra *commutativity axiom*

R9 $xy = yx$, for all $x, y \in S$.

Many of the most important and commonly occurring rings are commutative.

To summarize the main definition of this section: a ring is a set S, together with two binary operations $+$ and $.$, such that the ring axioms **R1–R8** hold. Axioms **R1–R4** say that S is an Abelian group with respect to the operation $+$ (one often refers to $(S, +)$ as the *additive group of the ring S*). Axioms **R5**, **R6** say that S is a (not necessarily commutative) semigroup with unit element with respect to the operation $.$ (one often refers to $(S, .)$ as the *multiplicative semigroup of the ring S*). Axioms **R7**, **R8** say that the two ring operations are connected by the distributive laws. It is usual to refer to $(S, +, .)$ simply as *the ring S*, i.e. to suppress explicit mention of the ring operations. The unit element 1 of S is often called the 'identity element' of S. The element 0 of S is usually called the 'zero element' of S.

Example 164. The set Z of all integers, with the usual operations of addition and multiplication, is a commutative ring. The ring axioms **R1–R8**, and also the commutativity axiom **R9**, hold by the laws of elementary arithmetic. In the same way, the set R of all real numbers, and the set Q of all rational numbers, are commutative rings.

Example 165. Rings of residue classes. Let n be a positive integer. We saw in section 2.5 that the relation 'congruence mod m' partitions Z into m *congruence classes* $E_0, E_1, \ldots, E_{m-1}$, which are often called *residue classes* mod m. In Example 118 (section 6.1) we saw that $E_0, E_1, \ldots, E_{m-1}$ are also the cosets of the subgroup mZ of Z (Z being regarded as the additive group $(Z, +)$), and in Example 136 (section 6.8) that these are the elements of the quotient group Z/mZ, which is an additive (Abelian) group, the rule for addition in Z/mZ being: $E_x + E_y = E_{x+y}$ for $x, y \in Z$. Now we may also define an operation of multiplication in Z/mZ by the rule: $E_x E_y = E_{xy}$ for $x, y \in Z$. With these rules, Z/mZ becomes a ring, called the *ring of residue classes* mod m. The ring axioms for Z/mZ follow immediately from the corresponding axioms for Z; notice that Z/mZ, like Z, is a commutative ring.

The addition and multiplication rules for Z/mZ are very simple, but contain a potential trap. We know that each class E_x may be written in many ways, in fact $E_x = E_{x'}$, for all x' such that $x \equiv x'$

mod m. Suppose that $E_x = E_{x'}$ and $E_y = E_{y'}$. Our rules give $E_x + E_y = E_{x+y}$ and also $E_{x'} + E_{y'} = E_{x'+y'}$, and so unless E_{x+y} is the same class as $E_{x'+y'}$, we should have an inconsistent (i.e. self-contradictory) rule for addition; similarly unless E_{xy} is the same class as $E_{x'y'}$, our rule for multiplication would be inconsistent. Fortunately all is well! For in section 2.6 we showed that from $x \equiv x'$ mod m and $y \equiv y'$ mod m there follow $x + y \equiv x' + y'$ and $xy \equiv x'y'$ mod m. Hence $E_{x+y} = E_{x'+y'}$ and $E_{xy} = E_{x'y'}$.

The tables below give the addition and multiplication in the ring $Z/6Z$. Note E_1 is the identity element, and E_0 the zero element, of this ring.

+	E_0	E_1	E_2	E_3	E_4	E_5
E_0	E_0	E_1	E_2	E_3	E_4	E_5
E_1	E_1	E_2	E_3	E_4	E_5	E_0
E_2	E_2	E_3	E_4	E_5	E_0	E_1
E_3	E_3	E_4	E_5	E_0	E_1	E_2
E_4	E_4	E_5	E_0	E_1	E_2	E_3
E_5	E_5	E_0	E_1	E_2	E_3	E_4

$\cdot$	E_0	E_1	E_2	E_3	E_4	E_5
E_0	E_0	E_0	E_0	E_0	E_0	E_0
E_1	E_0	E_1	E_2	E_3	E_4	E_5
E_2	E_0	E_2	E_4	E_0	E_2	E_4
E_3	E_0	E_3	E_0	E_3	E_0	E_3
E_4	E_0	E_4	E_2	E_0	E_4	E_2
E_5	E_0	E_5	E_4	E_3	E_2	E_1

The group $U(S)$ of invertible elements

An element x of a ring S is said to be *invertible*,† if there exists an element $y \in S$ such that $xy = yx = 1$. If such an element y exists, it is unique. For let $y' \in S$ be any element with the same property: $xy' = y'x = 1$. Then (using axioms **R5**, **R6**) we have $y = y1 = y(xy') = (yx)y' = 1y' = y'$. We therefore speak of y as *the* inverse of the invertible element x, and write $y = x^{-1}$.

THEOREM. *The set $U(S)$ of all invertible elements of a ring S, is closed to multiplication in S. $U(S)$ is a group, with multiplication as its binary operation.*

Proof. If x, x' are elements of $U(S)$, with inverses $y = x^{-1}$,

---

†Invertible elements of a ring S are sometimes called 'units' of S. This usage conflicts with our definition of units (section 4.3), and we shall avoid it.

$y' = x'^{-1}$, then $(xx')(y'y) = x(x'y')y = x1y = xy = 1$, and $(y'y)(xx') = y'(yx)x' = y'1x' = y'x' = 1$. Thus $xx' \in U(S)$, with $y'y$ as the inverse of xx'. This shows that $U(S)$ is closed to multiplication. To show that $U(S)$ is a group, we check that it satisfies the group axioms **G1**, **G2**, **G3** (section 4.4).

G1: $(xy)z = x(yz)$ holds for any x, y, $z \in U(S)$, since it holds for any x, y, $z \in S$ by **R5**. **G2**: The identity element 1 of S is invertible, with inverse $1^{-1} = 1$. So $1 \in U(S)$, and satisfies $1x = x1 = x$ for all $x \in U(S)$. **G3**: If $x \in U(S)$, then x^{-1} also belongs to $U(S)$, since $x^{-1}x = xx^{-1} = 1$ shows that x^{-1} has inverse x. Therefore $U(S)$ satisfies axiom **G3**.

Example 166. The only integers x which have inverse in Z, are $x = 1$ and $x = -1$. So $U(Z) = \{1, -1\}$. Notice that $U(Z)$ is *not* closed to addition. If R is the ring of real numbers, then every element x of R is invertible, except $x = 0$. So $U(R) = R - \{0\} = R^*$. The group $U(R)$ is the 'multiplicative group of the reals' which we met in Example 86 (section 4.5).

8.2 Elementary theorems on rings. Subrings

We give next some of the simplest theorems about rings. Like all theorems in ring theory, they are proved using only the ring axioms **R1**–**R8**. Let S be a ring. Our first theorem shows that the distributive laws **R7**, **R8** are still valid, if the operation $+$ is replaced by the operation of subtraction.

(i) THEOREM $(x - y)z = xz - yz$ *and* $x(y - z) = xy - xz$, *for all* x, y, $z \in S$.

Proof. By **R7** we have $(x - y)z + yz = [(x - y) + y]z = xz$. Add $-(yz)$ to both sides of this equation; we find $(x - y)z = xz - yz$. The proof that $x(y - z) = xy - xz$ is entirely similar, and we leave it to the reader.

(ii) THEOREM $x0 = 0 = 0x$, *for all* $x \in S$. *Therefore* 0 *is a zero element for the multiplicative semigroup* $(S, .)$.

Proof. Put $y = z = 0$ in the equation $x(y - z) = xy - xz$. We get $x(0 - 0) = x0 - x0$. Since $0 - 0 = 0$, and also

$x0 - x0 = x0 + (-x0) = 0$, this gives $x0 = 0$. The proof that $0x = 0$ is similar.

(iii) THEOREM $(-x)y = -xy = x(-y)$, *for all x, $y \in S$.*

Proof. Since $-x = 0 - x$ we have $(-x)y = (0-x)y = 0y - xy$ (using (i) above) $= 0 - xy$ (using (ii)) $= -xy$. The proof that $x(-y) = -xy$ is similar.

(iv) *General distributive laws*

   (a) $(x_1 + x_2 + \cdots + x_m)y = x_1 y + x_2 y + \cdots + x_m y$,
   (b) $x(y_1 + y_2 + \cdots + y_n) = xy_1 + xy_2 + \cdots + xy_n$,

and

   (c) $(x_1 + \cdots + x_m)(y_1 + \cdots + y_n) = \sum_{i=1}^{m} \sum_{j=1}^{n} x_i y_j$,

for all x, y, $x_1, \ldots, x_m$, $y_1, \ldots, y_n \in S$.

Proof. (a) We prove this by induction on m. Equation (a) holds trivially if $m = 1$. Suppose that $m \geq 2$, and that we have proved (a) for $m - 1$. We want to prove (a) for m. Put $u = x_1 + \cdots + x_{m-1}$ and $v = x_m$. Then $(u+v)y = uy + vy$ by axiom **R7**, hence $(x_1 + \cdots + x_{m-1} + x_m)y = (x_1 + \cdots + x_{m-1})y + x_m y$. Since (a) holds for $m-1$, we have $(x_1 + \cdots + x_{m-1})y = x_1 y + \cdots + x_{m-1} y$. Therefore $(x_1 + \cdots + x_{m-1} + x_m)y = x_1 y + \cdots + x_{m-1} y + x_m y$, which shows that (a) holds for m. This completes the proof of (a). We leave the proof of (b) to the reader. To prove (c), write $y = y_1, + \cdots + + y_n$ and apply (a), which shows that $(x_1 + \ldots + x_m)(y_1 + \cdots + y_n)$ is the sum of terms $x_i y$, for $i = 1, \ldots, m$. By (b) (taking $x = x_i$), we see that $x_i y = x_i y_1 + \cdots + x_i y_n$, i.e. it is the sum of the terms $x_i y_j$, for $j = 1, \ldots, n$. Therefore $(x_1 + \cdots + x_m)(y_1 + \cdots + y_n)$ is the sum of all the mn terms $x_i y_j$ obtained by giving i all values $1, \ldots, m$ and j all values $1, \ldots, n$; i.e. it is the sum $\sum_{i=1}^{m} \sum_{j=1}^{n} x_i y_j$.

Remark

Because addition in a ring is commutative, it does not matter in what order the terms $x_i y_j$ of this sum are given. But because multiplication is not assumed to be commutative, we may not in general replace $x_i y_j$ by $y_j x_i$, unless S is a commutative ring.

Example 167. Suppose that the identity and zero elements of a ring S coincide, $1 = 0$. Then S has only one element, $S = \{0\}$. For let x be any element of S. Then $x = x1$ by **R6**, so $x = x0$ because $1 = 0$. But $x0 = 0$ by Theorem (ii) above. Hence $x = 0$.

Example 168. Let x, y be elements in a ring S, and let m, n be *positive* integers. Put $x_1 = \cdots = x_m = x$ and $y_1 = \cdots = y_n = y$ in (iv)(c). We find that $(mx)(ny) = (mn)(xy)$, since all the terms in the sum $\Sigma\Sigma x_i y_j$ are equal to xy, and there are mn such terms. We can easily show that the equation $(mx)(ny) = (mn)(xy)$ holds for *all* integers m, n, including negative integers and zero. For example suppose m is negative and n is positive. Then $m = -p$, for some positive integer p. Thus $mx = -px$ (by the 'additive' version of Theorem (v), section 4.6), so $(mx)(ny) = (-px)(ny) = -[(px)(ny)]$ using Theorem (iii), and $(px)(ny) = (pn)(xy)$, because p, n are both positive. Thus $(mx)(ny) = -(pn)(xy) = (-pn)(xy) = (mn)(xy)$, where we use Theorem (iii) to prove the second equality. Similar arguments show that $(mx)(ny) = (mn)(xy)$ holds in all cases where m, n are both non-zero. Finally it holds whenever one or both of m, n is zero, by Theorem (ii).

Subrings

Let $S = (S, +, .)$ be a ring.

DEFINITION. A subset U of S is a *subring of S*, if it satisfies the following three conditions:

SR1 U contains the identity element 1 of S,
SR2 $x, y \in U \Rightarrow x - y \in U$,

and

SR3 $x, y \in U \Rightarrow xy \in U$.

This corresponds to the idea of a subgroup of a group (see section 5.2). In fact **SR2** says that a subring U of a ring S *is* a subgroup of S, when S is considered as additive group $(S, +)$ (see condition S in section 5.2). Condition **SR3** says that U must also be closed to the operation of multiplication.

The condition **SR1** is sometimes omitted from the definition of a subring, or replaced by the weaker condition that U should contain some element i which acts as a unit for multiplication *in* U, but is not necessarily the identity element 1 of S (see Example 170, below). However, condition **SR1** is more convenient to use, and is adequate for our purposes.

Notice that any subring U of S is itself a ring using the addition and multiplication operations which U 'inherits' from S (compare this with the theorem in section 5.2, that any subgroup of a group is itself a group). For the ring axioms **R1–R8** for U are automatically satisfied since they hold for S.

Example 169. Z is a subring of the ring R of all real numbers. The ring Q of all rational numbers is also a subring of R. The set $U = 2Z$ of all even integers satisfies **SR2** and **SR3**, but is not a subring of Z since it does not satisfy **SR1**. Notice that the subset $\{0\}$ of a ring S is *never* a subring of S, unless $S = \{0\}$. For if $U = \{0\}$ satisfies **SR1**, we must have $0 = 1$, which implies $S = 0$ by Example 167.

Example 170. The subset $U = E_0, E_2, E_4$ of the ring $Z/6Z$ (Example 165) satisfies **SR2** and **SR3**, but not **SR1** since it does not contain the identity element E_1 of $Z/6Z$. Therefore U is not a subring of $Z/6Z$. But U is itself a ring, with E_4 its identity element.

Example 171. If U, V are subrings of S, then $U \cap V$ is a subring of S. For if U, V both satisfy the conditions **SR1**, **SR2** and **SR3**, it is clear that $U \cap V$ satisfies them. More generally, if $\mathscr{S}$ is any set of subrings of S, then $\cap \mathscr{S}$ is a subring of S (compare this with Example 98, section 5.2).

Example 172. If x is any element of S, we denote by $\mathrm{rg}\{x\}$ the set of all elements $s \in S$ which can be expressed in the form

$$s = z_0 1 + z_1 x + z_2 x^2 + \cdots + z_r x^r, \tag{8.1}$$

where r is any integer ≥ 0, and $z_0, z_1, z_2, \ldots, z_r$ are any integers (negative integers and zero being allowed). Then $U = \mathrm{rg}\{x\}$ is a subring of S, and is called the *subring of S generated by x* (this corresponds to the cyclic subgroup gp$x$ of a group G generated by an element x of G, see section 5.4). To check that $U = \mathrm{rg}\{x\}$ is a subring, first notice that $s = 1$

belongs to rg$\{x\}$ (take $r = 0$ and $z_0 = 1$ in (8.1) above), i.e. rg$x$ satisfies **SR1**. It is easy to check that **SR2** and **SR3** hold for $U = $ rg$\{x\}$, since the sum and product of elements of form (8.1) have the same form. Notice that *any* subring V of S, which contains x, also contains all elements of the form (8.1), i.e. rg$\{x\}$ is the 'smallest' subring of S which contains x.

8.3   Integral domains

The ring $Z/6Z$ of Example 165 contains elements E_2, E_3, neither of which is zero, but whose product $E_2 E_3$ is equal to E_0, the zero element of $Z/6Z$. Such elements are called *zero-divisors*; to be precise, if x, y are non-zero elements of a ring S such that $xy = 0$, then x, y are called zero-divisors of S. A ring may have many such pairs (x, y) of zero divisors. But it is useful to have a special name for commutative rings which (like Z and R) do *not* have zero divisors.

DEFINITION. A ring Z is an *integral domain* if it satisfies the three conditions

ID1 $1 \neq 0$,
ID2 S has no zero divisors, i.e. if $xy = 0$ for elements x, $y \in S$, then either $x = 0$, or $y = 0$, or both,
ID3 S is commutative.

Remarks

The first condition $1 \neq 0$ is to avoid including the 'trivial' ring $S = \{0\}$ (see Example 167). Some authors do not require an integral domain to be commutative.

THEOREM. *Let S be an integral domain, and let a, x, y be elements of S. If $ax = ay$ and $a \neq 0$, then we must have $x = y$, i.e. we may 'cancel' a from both sides of the equation $ax = ay$, as in elementary algebra.*

Proof. From $ax = ay$ we deduce that $ax - ay = 0$, hence $a(x - y) = 0$, by section 8.2, Theorem (i). Since $a \neq 0$ and S has no zero-divisors, we must have $x - y = 0$, i.e. $x = y$.

RINGS AND FIELDS 119

Example 173. Z and R are integral domains, but the ring $Z/6Z$ is not. The theorem above fails for $S = Z/6Z$; for example $E_2 E_2 = E_2 E_5$ (see Example 165) and $E_2 \neq E_0$, yet we cannot cancel E_2 since $E_2 \neq E_5$.

Example 174. Let m be an integer, $m \geq 2$. Then *the ring Z/mZ is an integral domain if and only if m is a prime number.* For if m is *not* prime there exist positive integers x, y, both less than m, such that $m = xy$. Hence $E_x E_y = E_{xy} = E_m = E_0$, and since $0 < x, y < m$ we know that E_x, E_y are both not equal to E_0. Therefore Z/mZ has zero-divisors, and so it is not an integral domain. Now assume that m is a prime. If E_x, E_y satisfy $E_x E_y = E_0$, then $E_{xy} = E_0$, which implies that m divides xy. But since m is prime, this can happen only if either m divides x, or y, or both (see theorem (iii), section 8.13 for a proof of this fact). But then either $E_x = E_0$, or $E_y = E_0$, or both. So Z/mZ has no zero divisors. It is a commutative ring, and $E_1 \neq E_0$ (since $E_1 = E_0$ would imply $1 \equiv 0$ mod m, i.e. $m = 1$). Therefore Z/mZ satisfies conditions **ID1**, **ID2** and **ID3**, i.e. it is an integral domain.

8.4   Fields. Division rings

By the definition in section 8.1, a ring S is an algebraic structure in which 'addition' and 'multiplication' can be performed; and the ring axioms **R1–R8** are designed to ensure that these operations share many of the properties of addition and multiplication in ordinary arithmetic. 'Subtraction' is also defined in any ring S, since axioms **R1–R4** require that S is an Abelian group with respect to the operation $+$. But there is no analogue of 'division' in an arbitrary ring S, and it is *not* required that S should be a group with respect to multiplication. In fact S cannot be a group with respect to multiplication except in the trivial case $S = \{0\}$. For there would have to be an element 0^{-1} in S, such that $00^{-1} = 1$. But $00^{-1} = 0$ by theorem (ii), section 8.2. So we should have $1 = 0$, hence $S = \{0\}$ by Example 167.

However, it sometimes happens that every *non-zero* element $x \in S$ as a multiplicative inverse, i.e. that there is an element x^{-1} in S such that $xx^{-1} = 1 = x^{-1}x$ – for example this is true for the ring $S = R$ of real numbers. In such a ring, the set $S - \{0\}$ is a group with respect to multiplication i.e. $S - \{0\} = U(S)$, the group of invertible elements of S.

DEFINITION. Let F be a ring with identity and zero elements 1 and 0, respectively. Then F is a *division ring* if it satisfies the two conditions:

F1 $1 \neq 0$,

and

F2 For each element x of F, such that $x \neq 0$, there is an element x^{-1} such that $xx^{-1} = 1 = x^{-1}x$.

DEFINITION. A ring F is a *field* if it satisfies conditions **F1**, **F2** and also

F3 F is commutative.

Remark

1 We see from these definitions that a field is the same thing as a commutative division ring. Division rings are sometimes called 'skew fields'.
2 If $x(\neq 0)$ is an element of a division ring F, then the element x^{-1} given by **F2** is called the 'multiplicative inverse of x', or simply the 'inverse of x'.

THEOREM. *Let F be a division ring, and let x, y be elements of F. Then*

(i) $x \neq 0$, $y \neq 0 \Rightarrow xy \neq 0$. *Hence F has no zero-divisors.*
(ii) $x \neq 0 \Rightarrow x^{-1} \neq 0$.
(iii) *The set $F - \{0\}$ is a group with respect to the operation of multiplication in F.*

Proof. (i) Let x^{-1}, y^{-1} be the inverses of x, y respectively. Then $y^{-1}x^{-1}xy = y^{-1}1y = y^{-1}y = 1$. So if xy were equal to 0, we should have $1 = (y^{-1}x^{-1})(xy) = (y^{-1}x^{-1})0 = 0$ by theorem (ii), section 8.2. But this contradicts condition **F1**. So in fact $xy \neq 0$.
(ii) If x^{-1} were equal to 0, we should have $1 = xx^{-1} = x0 = 0$, again contradicting **F1**. So in fact $x^{-1} \neq 0$.
(iii) By **F2**, all elements of $F - \{0\}$ are invertible, hence

$U(F) = F - \{0\}$. Now (iii) follows from the theorem in section 8.1 (p. 113).

Remark

$F - \{0\}$ is often written F^* or $F^\times$, and referred to as the *multiplicative group of F*. Notice that this is defined only when F is a division ring (or a field, which is the special kind of division ring whose multiplicative group is Abelian).

Example 175. The most important fields are R (often called the field of real numbers), Q (the field of rational numbers) and C (the field of complex numbers; this is defined in Example 178 below). The only non-commutative division ring which we shall mention is the ring H of *quaternions* (see Example 179).

Example 176. Every field F is an integral domain. For F satisfies conditions **F1**, **F3**, which are the same as conditions **ID1**, **ID3** for an integral domain. It also satisfies **ID2**, by theorem (i), above. Hence F is an integral domain. But there are integral domains which are not fields, e.g. Z, which is not a field, since for example, the element $3 \in Z$ has no inverse in Z. Notice that Z is a subring of the field Q (and of E); in general, any subring G of a field F, is an integral domain, but is not necessarily a field. A subring G of a field F which *is* a field (this happens if G satisfies, as well as the subring conditions **SR1**, **SR2**, **SR3**, also the condition: $x \in G$, $x \neq 0 \Rightarrow x^{-1} \in G$), is called a *subfield* of F. For example, Q is a subfield of R.

Example 177. Suppose that S is an integral domain which is *finite*. Then S must be a field. Certainly S satisfies **F1**, **F3**, which are the same as the conditions **ID1**, **ID3**. We shall now prove that S satisfies **F2**. Let x be any non-zero element of S. Define a map $f: S \rightarrow S$ by the rule: $f(a) = xa$, for all $a \in S$. The theorem in section 8.3 shows that $f(a) = f(b) \Rightarrow xa = xb \Rightarrow a = b$, for any $a, b \in S$; in other words, f is injective. Then *because S is finite, f* must also be surjective. For let $S = \{a_1, \ldots, a_n\}$. The elements $f(a_1), \ldots, f(a_n)$ are distinct, therefore any element $b \in S$ is equal to $f(a_i)$ for some $a_i \in S$. In particular there is some $a \in S$ such that $f(a) = 1$, i.e. $xa = 1$. Taking $a = x^{-1}$, we see that **F2** is satisfied; hence S is a field.

In Example 174 (section 8.3) we saw that Z/pZ is an integral domain, whenever p is a prime number. So since $Z/pZ = \{E_0, \ldots, E_{p-1}\}$ is finite, it is actually a field. We shall give another proof of this fact in section 8.14.

Example 178. Complex numbers. For the benefit of readers who have not met complex numbers, we give the definition here. For a much more detailed discussion, see W. Ledermann, *Complex Numbers*, Routledge and Kegan Paul.

Complex numbers were invented to avoid the inconveniences caused by the fact that the real number field R contains no element i such that $i^2 = -1$. We simply invent a symbol i, which will be treated like a variable x in ordinary algebra, but with the property that i^2 can be replaced by -1 wherever it appears. A *complex number* is an expression $a_0 + ia_1$, where a_0, a_1 are any real numbers. Given such complex numbers $\alpha = a_0 + ia_1$, $\beta = b_0 + ib_1$, we will consider α and β to be equal, if and only if $a_0 = b_0$ *and* $a_1 = b_1$. Let C denote the set of all complex numbers. We define operations of addition and multiplication by the rules: for any α, $\beta \in C$ let

$$\alpha + \beta = (a_0 + b_0) + i(b_0 + b_1),$$

and

$$\alpha\beta = (a_0 b_0 - a_1 b_1) + i(a_0 b_1 + a_1 b_0).$$

It is not necessary to memorize the rule for multiplication, it follows from the convention that $i^2 = -1$. For we may multiply out $\alpha\beta = (a_0 + ia_1)(b_0 + ib_1) = a_0 b_0 + ia_1 b_0 + ia_0 b_1 + i^2 a_1 b_1$ in the ordinary way, then put $i^2 = -1$. Collecting together the terms with i, we arrive at $a_0 b_0 - a_1 b_1 + i(a_0 b_1 + a_1 b_0)$.

It is now possible to check that C is a commutative ring; we leave it to the reader to verify that the ring axioms **R1**–**R9** all hold. The zero element of C is $0 + i0$; this 'zero complex number' is usually written just as 0. The identity element is $1 + i0$; this is usually written as 1. In general, a complex number $a_0 + i0$ is written as a_0, and in this way *we regard any real number as a complex number*, i.e. R is regarded as a subset of C. Notice that our rules for addition and multiplication of complex numbers, reduce to the addition and multiplication of real numbers: if $\alpha = a_0 + i0$, $\beta = b_0 + i0$, then $\alpha + \beta = (a_0 + b_0) + i0$ and $\alpha\beta = a_0 b_0 + i0$. Since R contains the complex number $1 = 1 + i0$, it satisfies the conditions **SR1**, **SR2**, **SR3** which show that R is a subring of C.

Finally we must prove that C is a field. It satisfies **F1**, **F3**. We must show that it satisfies **F2**. Let $\alpha \in C$, $\alpha \neq 0$. To say $\alpha \neq 0$, is to say that $\alpha = a_0 + ia_1 \neq 0 + i0$, i.e. that not *both* a_0 and a_1 are zero. It follows that $a_0^2 + a_1^2 \neq 0$. It is now convenient to introduce the idea of the *conjugate* of a complex number $\alpha = a_0 + ia_1$: this is the complex number $\bar{\alpha} = a_0 - ia_1$. From our multiplication rule, $\bar{\alpha}\alpha = (a_0^2 + a_1^2) + i(a_0 a_1 - a_1 a_0) = a_0^2 + a_1^2$. It follows that $\alpha\beta = 1$, where

$$\beta = \frac{1}{a_0^2 + a_1^2} \quad \bar{\alpha} = \frac{a_0}{a_0^2 + a_1^2} - i\frac{a_1}{a_0^2 + a_1^2}.$$

Thus α has inverse $\alpha^{-1} = \beta$, hence C satisfies **F2**, as required.

Notice that if we apply this process to a non-zero real number $\alpha = a_0 + i0$, we get $\alpha^{-1} = 1/a_0 + i0 = a_0^{-1}$. It is clear that R is a subfield of the field C.

Example 179. Quaternions. Around 1843 W. R. Hamilton discovered a remarkable generalization of complex numbers, which he called *quaternions*. In place of the symbol i, Hamilton used three symbols i, j, k, which are to obey the following *multiplication rules*:

$$i^2 = j^2 = k^2 = -1, \quad ij = k = -ji, \quad jk = i = -kj, \quad ki = j = -ik.$$

A quaternion is by definition any expression of the form $\alpha = a_0 + ia_1 + ja_2 + ka_3$, where a_0, a_1, a_2, a_3 are real numbers. If $\beta = b_0 + ib_1 + jb_2 + kb_3$ is another such expression, we say that α is equal to β if and only if $a_0 = b_0, a_1 = b_1, a_2 = b_2$ and $a_3 = b_3$. Assuming now that α, β are any quaternions, define their sum to be $\alpha + \beta = (a_0 + b_0) + i(a_1 + b_1) + j(a_2 + b_2) + k(a_3 + b_3)$, and their product $\alpha\beta = (a_0 + ia_1 + ja_2 + ka_3)(b_0 + ib_1 + jb_2 + kb_3)$ is calculated by multiplying out this last expression, and then using the multiplication rules above to bring this to the form $c_0 + ic_1 + jc_2 + kc_3$. The reader may check that we get

$$c_0 = a_0b_0 - a_1b_1 - a_2b_2 - a_3b_3, \quad c_1 = a_0b_1 + a_1b_0 + a_2b_3 - a_3b_2,$$
$$c_2 = a_0b_2 + a_2b_0 + a_3b_1 - a_1b_3 \quad \text{and} \quad c_3 = a_0b_3 + a_3b_0 + a_1b_2 - a_2b_1.$$

It may now be checked that the set H of all quaternions, with these operations of addition and subtraction, is a ring.† The zero element is $0 + i0 + j0 + k0$, and the identity element is $1 + i0 + j0 + k0$. It is usual, when writing a quaternion, to omit any 'zero terms' $i0, j0$ and $k0$; thus the zero and identity quaternions are written simply 0 and 1, respectively. Notice that H is *not* commutative (for example $ij = -ji \neq ji$). But H is a *division ring*. It satisfies condition **F1**; we must show that it satisfies **F2**. Let $\alpha \in H$, $\alpha \neq 0$. This means that if $\alpha = a_0 + ia_1 + ja_2 + ka_3$, then not all of a_0, a_1, a_2, a_3 are zero; hence the real number $a_0^2 + a_1^2 + a_2^2 + a_3^2$ is not zero. Just as in the case of complex numbers, we define the *conjugate of* α to be the quaternion $\bar{\alpha} = a_0 - ia_1 - ja_2 - ka_3$, and find that $\alpha\bar{\alpha}$ and $\bar{\alpha}\alpha$ are both

---

†It is laborious, although quite straightforward, to check the ring axioms for H directly. We shall give an easier method in Chapter 9, Example 268, using matrices.

equal to $N + i0 + j0 + k0$, where $N = a_0^2 + a_1^2 + a_2^2 + a_3^2$. Therefore α has inverse element

$$\alpha^{-1} = \beta = \frac{1}{N}\bar{\alpha} = \frac{a_0}{N} - \frac{ia_1}{N} - \frac{ja_2}{N} - \frac{ka_3}{N}.$$

This shows that H satisfies **F2**, and so is a division ring. We can regard R and C as subfields of H.

8.5 Polynomials

We come now to a fundamental construction. Suppose that S is any commutative ring. We take a symbol X, which is *not* an element of S, and whose purpose is to serve as a 'variable' or 'indeterminate' in the following definition.

DEFINITION. A *polynomial in X with coefficients in S* is an expression of the form

$$f = f_0 + f_1 X + f_2 X^2 + \cdots + f_n X^n, \tag{8.2}$$

where n is any integer $n \geq 0$, and $f_0, f_1, f_2, \ldots, f_n$ are any elements of S (these need not be distinct). For each integer $r \in \{0, 1, \ldots, n\}$, f_r is called the *coefficient of X^r in f*. f_0 is sometimes called the 'constant term' of f.

Remarks

1 It is convenient to assume the coefficient f_r is defined for *all* integers $r \geq 0$; we simply take $f_r = 0$, for all $r > n$. This means that the polynomial f could be written

$$f = f_0 + f_1 X + \cdots f_n X^n + 0 X^{n+1} + 0 X^{n+1} + 0 X^{n+2} + \cdots, \tag{8.3}$$

a formally infinite sum, or more concisely as

$$f = \sum_{r \geq 0} f_r X^r \quad \text{or} \quad f = \Sigma f_r X^r. \tag{8.4}$$

But in using (8.3) or (8.4) to denote a polynomial, we must remember that, for each polynomial f, there is always some integer $n \geq 0$, such that $f_r = 0$ for all $r > n$; thus the 'infinite sum' involved is really finite.

2 It is usual to omit terms $f_r X^r$ for which $f_r = 0$. So, for example, $f = X - X^3$ is written instead of $f = 0 + 1X + 0X^2 + (-1)X^3$. With this convention, X, X^2, X^3, ... are all polynomials, e.g. X^3 is the polynomial $0 + 0X + 0X^2 + 1X^3$.

3 In the same way, any element $s \in S$ can be regarded as the polynomial $s + 0X + 0X^2 + \cdots$. In particular, the elements 1 and 0 of S can be regarded as polynomials if we want. We then speak of 'the polynomial 1', meaning $1 + 0X + 0X^2 + \cdots$ or 'the zero polynomial', meaning $0 + 0X + 0X^2 + \cdots$.

DEFINITIONS. Let $f = f_0 + f_1 X + f_2 X^2 + \cdots$ and $g = g_0 + g_1 X + g_2 X^2 + \cdots$ be polynomials in x with coefficients in S. Then we define

(a) $f = g \Leftrightarrow f_r = g_r$, for all $r \geq 0$. In particular $f = 0 \Leftrightarrow f_r = 0$, for all $r \geq 0$.

(b) $f + g = (f_0 + g_0) + (f_1 + g_1)X + (f_2 + g_2)X^2 + \cdots$, i.e. $f + g$ is defined to be the polynomial q such that $q_r = f_r + g_r$, for all $r \geq 0$.

(c) $fg = f_0 g_0 + (f_0 g_1 + f_1 g_0)X + (f_0 g_2 + f_1 g_1 + f_2 g_0)X^2 + \cdots$, i.e. fg is defined to be the polynomial p such that

$$p_r = f_0 g_r + f_1 g_{r-1} + \cdots + f_{r-1} g_1 + f_r g_0,$$

for all $r \geq 0$.

Remark

4 The definitions just given for $f + g$, fg conform to the usual practice in elementary algebra. For example, if we 'multiply out' the product $(f_0 + f_1 X + f_2 X^2 + \cdots)(g_0 + g_1 X + g_2 X^2 + \cdots)$, and then collect the terms involving each power X^r, we get the polynomial $p = p_0 + p_1 X + p_2 X^2 + \cdots$, as given by (c).

THEOREM. (i) *Let S be a commutative ring, and denote by $S[X]$ the set of all polynomials in X with coefficients in S. Then $S[X]$ is a commutative ring, using the operations of addition and multiplication defined in (b), (c) above. The polynomial 1 is the identity element of $S[X]$, and the zero polynomial is the zero element of $S[X]$.*

Proof. This consists in checking the ring axioms **R1–R9** for $S[X]$. Some of this is done in the examples below; the rest is left as an exercise for the reader.

Example 180. It is easy to check axioms **R1–R4** for $S[X]$. To check **R3**, for example, we must show that for each polynomial f in $S[X]$, there is a polynomial $-f$ in $S[X]$ such that $f + (-f) = 0 = (-f) + f$. It is clear that the polynomial $(-f_0) + (-f_1)X + (-f_2)X^2 + \cdots (-f_n)X^n$ has the required properties.

Example 181. To check **R5**, suppose $f, g, h \in S[X]$. We must prove, using the definition (c) for multiplication, that $(fg)h = f(gh)$. Let $a = (fg)h$ and $b = f(gh)$. In order to show that a and b are equal, we must show that $a_r = b_r$ for all $r \geq 0$ (we are using the definition (a) for equality of polynomials). To calculate the coefficient a_r of $a = (fg)h$, first let $u = fg$. Since $a = uh$, we have by definition (c)

$$a_r = u_0 h_r + u_1 h_{r-1} + \cdots + u_r h_0, \qquad \text{for all } r \geq 0. \tag{8.5}$$

We have also that

$$u_s = f_0 g_s + f_1 g_{s-1} + \cdots + f_s g_0, \qquad \text{for all } s \geq 0. \tag{8.6}$$

Putting these values for $u_0, \ldots, u_r$ into (8.5) we find

$$s_r = f_0 g_0 h_r + f_0 g_1 h_{r-1} + f_1 g_0 h_{r-1} + f_0 g_2 h_{r-2} + f_1 g_1 h_{r-2} + f_2 g_0 h_{r-2} + \cdots \tag{8.7}$$

On the right of equation (8.7) is a sum of products $f_i g_j h_k$, and each triple (i, j, k) appearing in (8.7) satisfies the conditions

$$i \geq 0, j \geq 0, k \geq 0 \qquad \text{and} \qquad i + j + k = r. \tag{8.8}$$

Conversely, let (i, j, k) be any triple satisfying (8.8). Let $s = i + j$. Then there is a term $u_s h_{r-s}$ in (8.5), and a term $f_i g_j$ in (8.6). So (8.7) must include a term $f_i g_j h_{r-s}$ (notice that $r - s = k$, because $i + j + k = r$). It is also clear that this product $f_i g_j h_k$ appears only once in (8.7). *So a_r is the sum of all products $f_i g_j h_k$, for which (i, j, k) satisfies (8.8).*

Now we make a similar calculation for $b = f(gh)$. We put $v = gh$, and calculate $b = fv$, using definition (c) to find the coefficients in the products bv and $v = gh$. The formula for v_s is

$$v_s = g_0 h_s + g_1 h_{s-1} + \cdots g_s h_0, \qquad \text{for all } s \geq 0,$$

and putting these values for $v_0, \ldots, v_r$ into

$$b_r = f_0 v_r + f_1 v_{r-1} + \cdots + f_r v_0,$$

we get

$$
\begin{aligned}
b_r = f_0 g_0 h_r + f_0 g_1 g_{r-1} + \cdots & \\
+ f_1 g_0 h_{r-1} + f_1 g_1 h_{r-2} + \cdots & \quad (8.9) \\
+ f_2 g_0 h_{r-2} + \cdots . &
\end{aligned}
$$

But it can be seen, in the same way as for our formula (8.7), that the right side of (8.9) is again exactly the sum of all $f_i g_j h_k$ for triples (i, j, k) which satisfy (8.8). Of course, these terms $f_i g_j h_k$ appear in a different order in (8.9), but this does not matter, because addition in S is commutative. Therefore $a_r = b_r$ for all $r \geq 0$, hence $a = b$, as required.

To check the commutativity axiom **R9** for $S[X]$, let $f, g \in S[X]$, and let $p = fg$ and $t = gf$. The coefficients of X^r in these polynomials are $p_r = f_0 g_r + \cdots + f_r g_0$ and $t_r = g_0 f_r + \cdots + g_r f_0$, respectively. Using the facts that both multiplication and addition are commutative in S, we have $p_r = t_r$, for all $r \geq 0$. Therefore $p = t$, as required.

Example 182. We can regard S as a subset of $S[X]$, by regarding each element $s \in S$ as a polynomial $s + 0X + 0X^2 + \cdots$ (Remark 3, above). Then S is a subring of $S[X]$ (check **SR1**, **SR2**, **SR3**).

Degree of a polynomial

Let $f = \Sigma f_r X^r = f_0 + f_1 X + \cdots$ be an element of $S[X]$, and let $f \neq 0$. Then the largest integer n such that $f_n \neq 0$, is called the *degree of f*, denoted $\deg f$. The term $f_n X^n$ is called the *leading term* (or highest term) of f.

Example 183. The polynomial $f = 3 + 2X^2 - 8X^3$ (regarded as element of $Z[X]$, or of $R[X]$) has degree 3, and leading term $-8X^3$.

A polynomial $f = f_0 + f_1 X + f_2 X^2 + \cdots$ of $S[X]$ has degree 0 if and only if $f_0 \neq 0$, and $f_1 = f_2 = \cdots = 0$. In other words, the elements of degree 0 in $S[X]$, are just the non-zero elements of S (regarding S

as subring of $S[X]$, see Example 182). Notice *that the polynomial 0 has no degree*, by our definition. (Some authors give the zero polynomial the degree $-\infty$, but we shall not dot this.)

The next theorem is very useful; notice, however, that it works only if S is an integral domain.

THEOREM. (ii) *Let S be an integral domain, and let f, g be elements of $S[X]$, with $f \neq 0$, $g \neq 0$. Then $fg \neq 0$, and*

$$\deg fg = \deg f + \deg g;$$

in fact the leading term of fg, is the product of the leading terms of f and g.

COROLLARY. *If S is an integral domain, then $S[X]$ is an integral domain.*

Proof of theorem. If $\deg f = m$ and $\deg g = n$, we have

$$f = f_0 + f_1 X + \cdots + f_m X^m, \qquad \text{with } f_m \neq 0,$$

and

$$g = g_0 + g_1 X + \cdots + g_n X^n, \qquad \text{with } g_n \neq 0.$$

We use definition (c) to calculate some of the coefficients of the polynomial $p = fg$. In particular $p_{m+n} = f_0 g_{m+n} + \cdots + f_{m-1} g_{n+1} + f_{m+1} g_{n-1} + \cdots + f_{m+n} g_0$. All the terms in this sum which precede $f_m g_n$ are zero, because $g_{m+n} = \cdots = g_{n+1} = 0$; similarly the terms which follow $f_m g_n$ are zero, because $f_{m+1} = \cdots = f_{m+n} = 0$. So $p_{m+n} = f_m g_n$, and this is non-zero, because S is an integral domain, and f_m, g_n are both non-zero. On the other hand if $r > m+n$, then $p_r = 0$, since p_r is a sum of terms $f_i g_j$ such that $i \geq 0, j \geq 0$ and $i+j = r$. For such a pair (i, j) either $i > m$, or $j > n$; in either case, $f_i g_j = 0$.

What we have just proved, shows that fg has degree $m+n$, and in fact that the leading terms of fg is $f_m g_n X^{m+n}$, which is the product of the leading terms $f_m X^m$ and $g_n X^n$ of f and g. Thus all parts of the theorem are proved. The corollary follows, since the theorem shows that if $fg = 0$, then either $f = 0$ or $g = 0$.

Example 184. If S is not an integral domain, then $S[X]$ is itself not an integral domain. For S must have a pair of non-zero elements, say x, y, such that $x \neq 0$, $y \neq 0$ and $xy = 0$. But we can regard x, y as polynomials of degree 0 (see Example 183) and as such, they are zero divisors in $S[X]$.

Example 185. Let f, $g \in S[X]$, with $f \neq 0$, $g \neq 0$. There is no exact formula for $\deg(f + g)$; the best we can say is that $\deg(f + g) \leq \max\{\deg f, \deg g\}$ – and of course even this has no meaning if $f + g = 0$. The reader will see that $\deg(f + g)$ can be equal to $\max\{\deg f, \deg g\}$ (take e.g. $f = 1 + X - X^2$, $g = 1 + 3X$). (Notation: if m, n are any real numbers, $\max\{m, n\}$ is defined to be m, if $m \geq n$, and to be n, if $m \leq n$.)

Example 186. Polynomials in several variables. For any commutative ring S, we define a ring $S[X, Y]$ as follows: first take the ring $S[X]$ of polynomials in a variable X, then let $S[X, Y]$ be the ring $(S[X])[Y]$ of polynomials in a new variable Y, with coefficients in $S[X]$. Such a polynomial

$$f = f_0(X) + f_1(X)Y + f_2(X)Y^2 + \cdots + f_n(X)Y^n,$$

where $f_0(X), f_1(X), \ldots$ are elements of $S[X]$, is called a *polynomial in X, Y with coefficients in S.* This polynomial can always be written as a (finite) sum of form

$$f = a_{00} + a_{10}X + a_{01}Y + a_{20}X^2 + a_{11}XY + a_{02}Y^2 + \cdots$$

where the a_{ij} are elements of S. The elements $X^i Y^j$ ($i \geq 0$, $j \geq 0$) are called *monomials* in X, Y (and by convention, $X^0 Y^0 = 1$); the element a_{ij} is called the *coefficient of $X^i Y^j$ in f.* One can extend these ideas to define polynomials in any number of variables.

8.6 Homomorphisms. Isomorphism of rings

In this section we return to the general theory of rings, and consider the idea of a homomorphism between two rings. This is exactly parallel to the idea of a homomorphism between groups (section 7.1).

DEFINITION. Let S, T be rings, with identity elements 1_S, 1_T respectively. Then a *homomorphism* of S into T is a map $\theta: S \to T$ which satisfies the following three conditions

RH1 $\theta(1_S) = 1_T$,
RH2 $\theta(x + y) = \theta(x) + \theta(y)$

and

RH3 $\theta(xy) = \theta(x)\theta(y)$,

for all $x, y \in S$.

Remarks

1 A homomorphism, in this sense, is often called a *ring-homomorphism*, or a *homomorphism of rings*, to distinguish it from a homomorphism of groups.

2 **RH2** shows that a ring-homomorphism $\theta: S \rightarrow T$ is a homomorphism between the additive groups $(S, +)$, $(T, +)$. Thus we may apply lemmas (i), (ii) of section 7.2, and deduce that $\theta(x - y)$ $\theta(x) - \theta(y)$ for all $x, y \in S$, and $\theta(0_S) = 0_T$, where $0_S, 0_T$ are the zero elements of S, T respectively (remember that these are the 'unit elements' of the additive groups $(S, +)$, $(T, +)$).

Monomorphisms, epimorphisms, etc. for rings

These are defined as follows. A ring-homomorphism $\theta: S \rightarrow T$ is called a *monomorphism* if θ is injective, an *epimorphism* if θ is surjective, and an *isomorphism* if θ is bijective. An *endomorphism* of S is a homomorphism $\theta: S \rightarrow S$ of the ring S into itself; if $\theta: S \rightarrow S$ is bijective, i.e. if θ is an isomorphism of S into itself, it is called an *automorphism* of S.

We sometimes speak of ring-monomorphisms, ring-isomorphism etc., if we wish to emphasize that we are dealing with homomorphisms between rings.

LEMMA (i) *If S, T, U are rings, and if $\theta: S \rightarrow T$, $\phi: T \rightarrow U$ are both ring-homomorphisms, then $\phi\theta: S \rightarrow U$ is a ring-homomorphism. If θ and ϕ are both ring-isomorphisms, then $\phi\theta$ is a ring-isomorphism.*
(ii) *If $\theta: S \rightarrow T$ is a ring-isomorphism, then $\theta^{-1}: T \rightarrow S$ is a ring-isomorphism.*
These lemmas correspond to lemmas (iv), (v) of section 7.2, and

they can be proved in much the same way. We leave the proofs of lemmas (i), (ii) as an exercise for the reader.

DEFINITION. Let S, T be rings. Then we say that S and T are *isomorphic* and write $S \cong T$, if there exists an isomorphism $\theta : S \to T$.

Isomorphism for rings, is the analogue of isomorphism for groups. Rings which are isomorphic are 'equivalent' from the point of view of ring theory. The relation $\cong$ has the properties of an equivalence relation (see section 7.3).

Example 187. Let m be a positive integer, and let $Z/mZ = \{E_0, E_1, \ldots, E_{m-1}\}$ be the ring of residue-classes mod m (see Example 165, section 8.1). Then the map $\theta : Z \to Z/mZ$, which takes each $x \in Z$ to its residue-class E_x, is a ring-homomorphism. For we have **RH1**: $\theta(1) = E_1$, which is the identity element of Z/mZ, **RH2**: $\theta(x+y) = E_x + E_y = \theta(x) + \theta(y)$, and **RH3**: $\theta(xy) = E_{xy} = E_x E_y = \theta(x)\theta(y)$, for all x, $y \in Z$. Since θ is surjective, it is an epimorphism.

Example 188. Characteristic map of a ring. Let S be any ring, with identity element 1_S. The map $\gamma : Z \to S$ given by the rule $\gamma(n) = n1_S$, for all $n \in Z$, is called the characteristic map of S. It is a ring-homomorphism, for we have **RH1**: $\gamma(1) = 1.1_S = 1_S$, **RH2**: $\gamma(m+n) = (m+n)1_S = m1_S + n1_S = \gamma(m) + \gamma(n)$ (by the additive version of the general index law for groups, Example 91, section 4.6: this says $mx + nx = (m+n)x$, for x in any additive group), and finally **RH3**: $\gamma(mn) = (mn)1_S = (m1_S)(n1_S) = \gamma(m)\gamma(n)$ (by taking $x = y = 1_S$ in Example 168, section 8.2). In general, γ is neither injective nor surjective.

Example 189. Let $\alpha = a_0 + a_1 i$ be an element of the field C (Example 178, section 8.4), then we have defined the conjugate of α to be the complex number $\bar{\alpha} = a_0 - a_1 i$. It is easy to check that $\overline{\alpha + \beta} = \bar{\alpha} + \bar{\beta}$ and $\overline{\alpha\beta} = \bar{\alpha} . \bar{\beta}$, for any α, $\beta \in C$. It is also clear that $\bar{1} = 1$ (since the complex number 1 is, by definition, $1 + 0i$). Therefore the map $\pi : C \to C$ given by $\pi(\alpha) = \bar{\alpha}$ for all $\alpha \in C$, is a ring-homomorphism. It is in fact an *automorphism* of C, as the reader can check, by showing that π is bijective.

Example 190. A property of rings is said to be *invariant*, if whenever S, T are isomorphic rings and S has the property, then also T has it. For example if S is commutative and $S \cong T$, then also T is commutative. For let

u, $v \in T$, and let $\theta: S \to T$ be an isomorphism. If $x = \theta^{-1}(u)$, $y = \theta^{-1}(v)$, then x, $y \in S$ and hence $xy = yx$ because S is commutative. Apply θ to this equation. We get $\theta(x)\theta(y) = \theta(y)\theta(x)$, i.e. $uv = vu$. So T is commutative; this shows that the property of being commutative is invariant. In the same style, one can show that the property of being an integral domain, or of being a field, is invariant.

Example 191. To prove that the fields R and C are not isomorphic, i.e that they are not isomorphic as rings. Suppose we had a ring isomorphism $\theta: C \to R$. Let $s = \theta(i)$. Since $i^2 + 1 = 0$ in C, we have $\theta(i^2 + 1) = \theta(0) = 0$ in R; and $\theta(i^2 + 1) = \theta(i^2) + (1) = \theta(i)^2 + 1 = s^2 + 1$. But there is no real number s such that $s^2 + 1 = 0$. Therefore there is no isomorphism $\theta: C \to R$. In fact, the argument just given shows that there is no ring-homomorphism $\theta: C \to R$.

8.7   Ideals

Ideals are subsets of a ring S (the definition is given below) which play the same part as the normal subgroups of a group G. Normal subgroups are important in group theory for two main reasons, (i) if N is a normal subgroup of G, one can make a new group, the quotient group G/N (see section 6.8), and (ii) the kernel of a homomorphism $\theta: G \to H$ is always a normal subgroup of G. These two facts combine in the Homomorphism Theorem (section 7.4). We shall find exact parallels in ring theory to all of this.

DEFINITION. Let S be a ring. Then a non-empty subset J of S is called an *ideal of S* if it satisfies the conditions

I1   If x, $y \in J$, then $x - y \in J$,

and

I2   If $x \in J$ and $a \in S$, then ax and xa both belong to J.

Remarks

1   Condition **I1** says that an ideal J is a subgroup of the additive group $(S, +)$ (see section 5.2, condition **S**).
2   It is essential to realize that in condition **I2**, the element *a may be any element of S*; it does not have to be an element of J.

3 We shall see in the next example that if an ideal J contains the identity element 1 of S, then $J = S$. For this reason, an ideal J of S is never a subring of S, unless $J = S$. For a subring of S always contains 1, by **SR1** (section 8.2).

Example 192. The sets $\{0\}$ and S are always ideals of S. If an ideal J of S contains the identity element 1 of S, then taking $x = 1$ in **I2** we find that J contains $a1 = a$, for all $a \in S$. Hence $J = S$.

DEFINITION. A ring S is *simple* provided $1 \neq 0$ (so that S is not the trivial ring $\{0\}$, see Example 167, section 8.2), and S has no ideals except $\{0\}$ and S. (This corresponds to the idea of a simple group – see Example 133, section 6.7.)

Example 193. Any division ring (and in particular any field) is simple. For let F be a division ring. We know that $1 \neq 0$ in F, by condition **F1** (section 8.4). Now let J be any ideal of F, which is not the ideal $\{0\}$. J must contain a non-zero element x. But, taking $a = x^{-1}$ in **I2**, we see that J contains $x^{-1}x = 1$. So $J = F$, by Example 192. Therefore the only ideals of F are $\{0\}$ and F, hence F is simple. We shall see in the next chapter (section 9.10) that there are simple, non-commutative rings which are not division rings. But any simple, *commutative* ring is a field (see Exercise 21 at the end of this chapter).

Sum and intersection of ideals

If S is any ring, let $\mathscr{I}(S)$ be the set of all ideals of S. Of course if S is simple, then $\mathscr{I}(S) = \{S, \{0\}\}$, which is not very interesting. But often $\mathscr{I}(S)$ contains some useful information about S. This is partly due to the fact that there are some binary operations on $\mathscr{I}(S)$, as we shall now see.

LEMMA. *Let J, K be ideals in a ring S. Then the sets (i) $J + K = \{u + v \mid u \in J, v \in K\}$ and (ii) $J \cap K$ are both ideals of S.*

Proof. (i) Let $x = u + v$, $y = u' + v'$ be elements of $J + K$, with u, $u' \in J$ and v, $v' \in K$. Then $x - y = (u - u') + (v - v') \in J + K$, since $u - u' \in J$, $v - v' \in K$ (using **I1** for J, K respectively); hence $J + K$ satisfies condition **I1**. If a is any element of S, then $ax = a(u+v) = au + av \in J+K$, since $au \in J$, $av \in K$ (using **I2** for J, K respectively); similarly $xa \in J+K$. Hence $J+K$ satisfies condition **I2**, and is therefore an ideal of S.

(ii) Let $x, y \in J \cap K$. Then $x - y \in J$ (by **I1** for J) and $x - y \in K$ (by **I1** for K); hence $x - y \in J \cap K$. Similarly if a is any element of S, then ax, xa both belong to J, and also to K; hence $ax, xa \in J \cap K$. This proves that $J \cap K$ is an ideal of S.

From this lemma it follows that there exist binary operations 'sum' and 'intersection' on $\mathscr{I}(S)$, which take any given pair (J, K) of ideals of S, to the ideals $J+K$ and $J \cap K$, respectively. (There is another binary operation on $\mathscr{I}(S)$, which takes (J, K) to the 'product' JK, see Example 196, below. This is important for more advanced ring theory, but we shall not need it in this book.)

Example 194. The operations 'sum' and 'intersection' are both commutative and associative. As far as intersection is concerned, this follows from Examples 67 and 68 (section 4.2). That $J+K = K+J$, is clear from the definition of $J+K$ and the fact that addition in S is commutative. If J, K, L are ideals of S, then $(J+K)+L = J+(K+L)$, using the associativity of addition in S. Note that $J+J = J$ and $J \cap J = J$, for any $J \in \mathscr{I}(S)$.

Example 195. In general, the union $J \cup K$ of ideals J, K is not an ideal (it does not satisfy **I1**). However, $J+K$ contains $J \cup K$, and any ideal L which contains both J and K, must contain also all elements $u+v$ ($u \in J$, $v \in K$), hence $L \supseteq J+K$. So $J+K$ is the 'smallest' ideal of S, which contains both J and K.

Example 196. If J, K are ideals of S, the 'product' JK is, by definition, the set of all elements x of S which can be written in the form $x = u_1 v_1 + \cdots + u_r v_r$, for any $r \geq 1$, and any elements $u_1, \ldots, u_r \in J$ and $v_1, \ldots, v_r \in K$. It is quite easy to check that JK is an ideal of S. In general $JK \neq KJ$.

8.8   Quotient rings

Let S be a ring, and let J be an ideal of S. By condition **I1** (see the beginning of section 8.7) J is a subgroup of $(S, +)$. So we may define the quotient *group* S/J as in section 6.8; S/J is the set of all (additively written) cosets $a+J=\{a+x\,|\,x\in J\}$, with operation of *addition* defined on S/J by the rule

$$(a+J)+(b+J)=(a+b)+J, \tag{8.10}$$

for all a, $b\in S$. Now we shall define an operation of *multiplication* on S/J by the rule

$$(a+J)(b+J)=ab+J, \tag{8.11}$$

for all a, $b\in S$. We must show that this rule (2) is consistent, i.e. if a, a', b, $b'\in S$ are such that $a+J=a'+J$ and $b+J=b'+J$, we must prove that $ab+J=a'b'+J$. Here we shall need the fact that J satisfies condition **I2**. We have $ab-a'b'=ab-a'b+a'b-a'b'=(a-a')b+a'(b-b')$. Since $a+J=a'+J$ we have $a-a'\in J$. By **I2**, this implies that $(a-a')b\in J$. Similarly $b-b'\in J$, so by **I2** this implies $a'(b-b')\in J$. Since J is closed to addition, we have $ab-a'b'=(a-a')b+a'(b-b')\in J$, i.e. $ab+J=a'b'+J$, which is what we wanted to prove. We are now in a position to state the following, which is the analogue of the theorem in section 7.8.

THEOREM. *Let J be an ideal of a ring S, and let operations of addition and multiplication be defined on S/J by the rules (8.10) and (8.11). Then S/J is a ring. The identity and zero elements of S/J are $1+J$ and $0+J$, respectively.*

Proof. The ring axioms **R1–R8** for S/J follow very easily from the corresponding axioms for S. For example, to show that S/J satisfies the distributive axiom **R7**, we take elements $x=a+J$, $y=b+J$, $z=c+J$ of S/J (so that a, b, c are elements of S). By our rules (8.10) and (8.11) we find that $(x+y)z$ is equal to

$$[(a+J)+(b+J)](c+J)=[(a+b)+J](c+J)=(a+b)c+J,$$

while $xz+yz$ is equal to

$$(a+J)(c+J)+(b+J)(c+J)=(ac+J)+(bc+J)=(ac+bc)+J.$$

But $(a+b)c=ac+bc$, since **R7** holds for S. Hence the two expressions above are equal. This shows that **R7** holds for S/J. The reader can check the other axioms for S/J in a similar way; in checking **R3** (existence of 'negative') we must take $-(a+J)$ to be the coset $(-a)+J$, because $(a+J)+[(-a)+J]=$ $(a+(-a))+J=0+J$, for any element $a=J$ of S/J.

DEFINITION. Let J be an ideal in a ring S. Then the ring S/J is called the *quotient ring*† (or *factor ring*) of S by J.

Notation for quotient rings

The 'coset' notation $a+J$ for elements of S/J is inconvenient, and one often uses other notations such as $[a]$ or $\bar{a}$. We shall usually write $[a]$ for $a+J$. Then the rules (8.10) and (8.11) become

$$[a]+[b]=[a+b] \quad and \quad [a][b]=[ab] \quad for \ all$$
$$[a], [b] \ in \ S/J, \tag{8.12}$$

but it is essential to remember also the *equality rule*

$$[a]=[b]\Leftrightarrow a+J=b+J\Leftrightarrow a-b\in J. \tag{8.13}$$

Example 197. We have already met an example of a quotient ring in Example 165 (section 8.1). If m is a positive integer, it is easy to check that mZ is an ideal of Z (this is an example of a principal ideal, see section 8.10). The class E_x is the same as the coset $x+mZ$, and the rules for adding and multiplying such cosets E_x, E_y (see Example 165) are exactly the rules (8.10) and (8.11) given above. So the ring of residue classes mod m is the same as the quotient ring Z/mZ.

Example 198. If S is commutative, then so is S/J, for any ideal J of S. For if $a+J=[a]$, $b+J=[b]$ are any elements of S/J we have (using (8.12) above) $[a][b]=[ab]=[ba]=[b][a]$.

Example 199. We saw in Example 177 (section 8.4) that Z/pZ is a field, if p is a prime number. Using the notation $[a]$ (instead of E_a) for an element of Z/pZ, we know that every non-zero element $[a]$ of Z/pZ has an inverse, say $[a]^{-1}=[b]$. Of course we cannot take $b=a^{-1}$, since a^{-1} is usually not an element of Z. Instead we find an integer b, such that $[a][b]=[1]$, i.e. such that $[ab]=[1]$ (using (8.12)), i.e. such that $ab-1\in pZ$ (using the

†By an unfortunate coincidence, the term 'quotient ring' is sometimes used in a different sense, to mean 'ring of quotients' or 'ring of fractions'.

equality rule (8.13) with $J = pZ$), i.e. such that $ab \equiv 1 \bmod p$. We shall give in section 8.14 a systematic method of calculating such an integer b. But for a small prime p, it is easier to find the inverse of $[a]$ by trial: for example, if $p = 7$, and we want the inverse of $[3]$, we can just multiply $[3]$ by each of the elements $[1], [2] \ldots, [6]$ until we find some $[b]$ for which $[3][b] = [1]$, i.e. such that $3b \equiv 1 \bmod 7$. We find $[3][5] = [15] = [1]$, so that $[3]^{-1} = [5]$.

Example 200. Let $S = R[X]$, the ring of polynomials in X with real numbers as coefficients. Then $X^2 + 1$ is an element of S, and we consider the set $J = \{q \cdot (X^2 + 1) | q \in S\}$, i.e. J is the set of all polynomials which are divisible by $X^2 + 1$. It is easy to check that J is an ideal of S (see section 8.10).

We shall show that the quotient ring S/J is isomorphic to the field C of complex numbers. Write $\xi = [X] (= X + J)$. This is an element of S/J, and we have $\xi^2 = [X][X] = [X^2]$. But from this we have $\xi^2 + [1] = [X^2] + [1] = [X^2 + 1] = [0]$; the last equality comes because $X^2 + 1 \in J$. In other words, we have in our quotient ring S/J an element ξ such that $\xi^2 = -[1]$.

Now take any element $f + J$ in S/J. If $f = f_0 + f_1 X + \cdots + f_n X^n$, then $f + J = [f_0] + [f_1][X] + \cdots + [f_n][X^n] = [f_0] + [f_1]\xi + \cdots + [f_n]\xi^n$. Since $\xi^2 = -[1]$, we have $\xi^3 = -\xi$, $\xi^4 = [1]$, $\xi^5 = \xi$, etc.: it follows that $f + J$ can be written as $f + J = [a_0] + [a_1]\xi$, for suitable $a_0, a_1 \in R$. Define a map $\theta: C \to S/J$ by the rule: $\theta(a_0 + a_1 i) = [a_0] + [a_1]\xi$, for all complex numbers $a_0 + a_1 i$ (a_0, a_1 are elements of R). What we have just proved, shows that θ *is surjective.* If $\alpha = a_0 + a_1 i$ and $\beta = b_0 + b_1 i$ are any elements of C such that $\theta(\alpha) = \theta(\beta)$, then $[a_0] + [a_1]\xi = [b_0] + [b_1]\xi$, i.e. $[a_0 + a_1 X] = [b_0 + b_1 X]$. Then the equality rule (8.13) says that $(a_0 + a_1 X) - (b_0 + b_1 X)$ lies in J, i.e. that $(a_0 - b_0) + (a_1 - b_1)X = q \cdot (X^2 + 1)$ for some $q \in S = R[X]$. The polynomial on the left of this equality, *if it is not zero*, has degree 0 or 1. But the degree of $q \cdot (X^2 + 1) = \deg q + \deg X^2 = \deg q + 2$ (theorem (ii), section 8.5). This gives a contradiction. It follows that $(a_0 - b_0) + (a_1 - b_1)X = 0$, i.e. that $a_0 = b_0$ and $a_1 = b_1$. We have now proved that $\theta(\alpha) = \theta(\beta) \Rightarrow \alpha = \beta$; hence θ is *injective.* Finally we prove that $\theta: C \to S/J$ is a ring-homomorphism, by checking that it satisfies conditions **RH1, RH2, RH3** of section 8.6. For example if $\alpha = a_0 + a_1 i$, $\beta = b_0 + b_1 i$ we have by the multiplication rule for complex numbers $\alpha\beta = (a_0 b_0 - a_1 b_1) + (a_0 b_1 + a_1 b_0)i$ (see Example 178, section 8.4). Thus $\theta(\alpha\beta) = [a_0 b_0 - a_1 b_1] + [a_0 b_1 + a_1 b_0]\xi$. But since $\xi^2 = -1$, we find easily that this last expression is equal to $\theta(\alpha)\theta(\beta)$. This shows that θ satisfies **RH3**. We leave it to the reader to check that θ satisfies **RH1, RH2**. Thus θ is a ring-isomorphism, which shows that $C \cong S/J$.

8.9 The Homomorphism Theorem for rings

This section contains the analogue for rings, of the Homomorphism Theorem for groups (section 7.4). Let S, T be rings with identity elements 1_S, 1_T, and zero elements 0_S, 0_T, respectively.

DEFINITION. Let S, T be rings and $\theta: S \to T$ a homomorphism. The sets

$$\text{Ker } \theta = \{x \in S | \theta(x) = 0_T\}$$

and

$$\text{Im } \theta = \{\theta(x) | x \in S\}$$

are called the *kernel* and *image* of θ, respectively.

Notice that Ker θ is a subset of S, and Im θ is a subset of T.

THE HOMOMORPHISM THEOREM FOR RINGS. *Let S, T be rings and $\theta: S \to T$ a homomorphism. Then*

   (i) *Ker θ is an ideal of S.*
   (ii) *Im θ is a subring of T.*
   (iii) *If we write $J = Ker\ \theta$, there is an isomorphism $\theta^*: S/J \to Im\ \theta$,*

defined by the rule: if $x \in S$, then $\theta^*([x]) = \theta(x)$. Here $[x]$ denotes the coset $x + J$. It follows that $S/Ker\ \theta \cong Im\ \theta$.

Proof. By our assumption, $\theta: S \to T$ satisfies the conditions **RH1**, **RH2**, **RH3** (section 8.6). In particular, by **RH2**, θ is a homomorphism of the additive groups $(S, +) \to (T, +)$. Thus (i) Ker θ is a subgroup of $(S, +)$, i.e. it satisfies **I1** (section 8.7). Now let $x \in \text{Ker } \theta$ and $a \in S$. We have $\theta(ax) = \theta(a)\theta(x) = \theta(a)0_T = 0_T$, hence $ax \in \text{Ker } \theta$. Similarly $xa \in \text{Ker } \theta$. Therefore Ker θ satisfies condition **I2**, and so it is an ideal of S. (ii) Since $\theta(1_S) = 1_T$ by **RH1**, we know that Im θ contains 1_T, i.e. Im θ satisfies condition **SR1** (section 8.2). Also Im θ is a subgroup of $(T, +)$ because θ is a (group-)homomorphism $(S, +) \to (T, +)$; hence Im θ satisfies **SR2**. Finally Im θ satisfies **SR3**, since if $u, v \in \text{Im } \theta$, there are elements $x, y \in S$ such that $\theta(x) = u$, $\theta(y) = v$. Thus $\theta(xy) = \theta(x)\theta(y) = uv$, and this shows that $uv \in \text{Im } \theta$. Therefore Im θ is a subring of T. (iii) From the Homomorphism Theorem for groups (regarding θ as homomorphism $(S, +) \rightarrow (T, +)$ once more), we know that θ^* is well defined, is bijective, and satisfies **RH2**. So it remains to show that θ^* satisfies **RH1**, **RH3**. It satisfies **RH1** because $\theta^*([1_S]) = \theta(1_S) = 1_T$, and θ^* satisfies **RH3** since $\theta^*([x][y]) = \theta^*([xy]) = \theta(xy) = \theta(x)\theta(y) = \theta^*([x])\theta^*([y])$, *for all* $x, y \in S$. This proves that θ^* is a ring-isomorphism. The proof of the theorem is complete.

Example 201. Natural epimorphisms. Let J be any ideal of a ring S. Define the map $v : S \rightarrow S/J$ by the rule: if $a \in S$, let $v(a) = [a] \, (= a + J)$. By the definition of S/J (section 8.8), we see that v is a ring-epimorphism: v is called the *natural epimorphism of S onto S/J*. Moreover Ker $v = J$, since for any $a \in S$, we have $a \in \text{Ker } v \Leftrightarrow v(a) = [0] \Leftrightarrow [a] = [0] \Leftrightarrow a \in J$, the last equality coming from the equality rule (8.13) (p. 136).

Example 202. If the homomorphism $\theta : S \rightarrow T$ is injective (i.e. if it is a monomorphism), then Ker $\theta = \{0_S\}$. For if $a \in \text{Ker } \theta$, we have $\theta(a) = 0_T = \theta(0_S)$, hence $a = 0_S$ because θ is injective. Conversely if $\theta : S \rightarrow T$ is any ring-homomorphism such that Ker $\theta = \{0_S\}$, then θ is injective. For if $a, b \in S$ then $\theta(a) = \theta(b) \Rightarrow \theta(a - b) = 0_T \Rightarrow a - b \in \text{Ker } \theta \Rightarrow a = b$ since Ker $\theta = \{0_S\}$.

8.10   Principal ideals in a commutative ring

Throughout this section S denotes a *commutative* ring. If b is a given element of S, we define $Sb = \{sb | s \in S\}$; this is a subset of S. Because S is commutative, Sb is the same as the set $bS = \{bs | s \in S\}$.

LEMMA. (i) *If b is any element of a commutative ring S, then the set $Sb = bS$ is an ideal of S.*

Proof. Sb satisfies conditions **I1**, since if $x, y \in Sb$ there are elements s, t of S such that $x = sb$, $y = tb$; hence $x - y = sb - tb = (s - t)b$, which belong to Sb. Now let a be any element of S, and $x = sb$ as before. Then $ax = xa = (as)b$ lies in Sb. Thus Sb satisfies **I2**, and so Sb is an ideal of S.

DEFINITION. If b is any element of a commutative ring S, then the ideal Sb of S is called the *principal ideal of S generated by b.*

Principal ideals are the easiest ideals to construct. More complicated ideals can be made by adding principal ideals; for example if b_1, b_2 are any elements of S, then $Sb_1 + Sb_2$ is an ideal of S (see the lemma in section 8.7). Generally, if $B = \{b_1, \ldots, b_n\}$ is any finite subset of S, we have an ideal $Sb_1 + \cdots + Sb_n$ called the *ideal of S generated by B*. It consists of all elements $s_1 b_1 + \cdots + s_n b_n$, where $s_1, \ldots, s_n$ are any elements of S.

Example 203. In general, the ideal $Sb_1 + \cdots + Sb_n$ is not principal if $n \geq 2$. But it may happen in special cases that such an ideal is principal. For example, take $S = Z$, and consider the ideal $J = 20Z + 12Z$, which consists of all integers $20z_1 + 12z_2$ with $z_1, z_2 \in Z$. Then J contains $20 \cdot (-1) + 12 \cdot 2 = 4$, hence (by property **I2**) J contains all elements $4z$, $z \in Z$; thus $J \supseteq 4Z$. But every element $20z_1 + 12z_2$ of J belongs to $4Z$, since $20z_1 + 12z_2 = 4(5z_1 + 3z_2)$. Hence $J = 4Z$, and so $J = 20z + 12z$ is equal to the principal ideal $4Z$. In fact the next theorem shows that *every* ideal of Z is principal.

THEOREM. *Every ideal J of Z is a principal ideal. The set of all ideals of Z is*

$$\mathscr{I}(Z) = \{0Z = \{0\}, 1Z = Z, 2Z, 3Z, \ldots \}.$$

Proof. Let J be any ideal of Z. By **I1**, J must be a subgroup of $(Z, +)$. But we proved in Example 154 (section 7.5) that every subgroup of $(Z, +)$ has the form mZ, for some $m \geq 0$; thus J belongs to the set $\mathscr{I}(Z)$ described above. Notice that the principal ideals $(-1)Z, (-2)Z, \ldots$ are not mentioned in $\mathscr{I}(Z)$, since for any $m \in Z$ there holds $mZ = \{\ldots, -2m, -m, 0, m, 2m, \ldots\} = (-m)Z$.

DEFINITION. A commutative ring S, such that every ideal of S is principal, is called a *principal ideal ring*. If S is also an integral domain, it is called a *principal ideal domain*, often written PID for short.

Since Z is an integral domain, the above theorem says that Z is a PID. It is a remarkable fact, which we shall prove in section 8.12, that the ring $F[X]$ of polynomials in X over a field F is also a PID.

Example 204. Characteristic of a ring. Let S be any ring, and let $\gamma: Z \to S$ be the 'characteristic map' defined in Example 188 (section 8.6). By the Homomorphism Theorem, Ker γ is an ideal of Z, so the theorem above says that Ker $\gamma = mZ$, for some integer $m \geq 0$. This integer is called the *characteristic of S*. By the Homomorphism Theorem again, Im $\gamma = \{n1_S | n \in Z\}$ is a subring of S, and $Z/mZ \cong \text{Im } \gamma$. If $m = 0$, the map $\gamma: Z \to S$ is a monomorphism, and the elements $\ldots, -31_S, -21_S, 0_S, 1_S, 21_S, 31_S, \ldots$ are distinct. If $m > 0$, then Im $\gamma = \{0_S, 1_S, 21_S, \ldots, (m-1)1_S\}$. If S is an integral domain (and in particular if S is a *field*) then the characteristic m of S is either 0, or it is a prime number. For $Z/mZ \cong \text{Im } \gamma$ must itself be an integral domain, and if $m > 0$ this happens only if m is a prime (Example 173, section 8.3).

Example 205. Let $S = Z[X]$, the ring of polynomials in X with coefficients in Z. Let $J = S.2 + S.X$, the ideal of S generated by the set $\{2, X\}$. We shall prove that J is *not* a principal ideal of S. For suppose it were, then $J = Sb$, for some element $b = b_0 + b_1 X + \ldots + b_n X^n$ in S. Since J contains the element 2 (regarding 2 as polynomial of degree 0), there must hold $2 = fb$, for some $f \in S$. Then $0 = \deg 2 = \deg f + \deg b$ by theorem (ii), section 8.5 (note that Z is an integral domain). But since $\deg f$ and $\deg b$ are both integers ≥ 0, we must have $\deg f = \deg b = 0$. Therefore $b = b_0$. Since $b \in S.2 + S.X$, there must exist $s, t \in S$ such that $b = b_0 = s.2 + t.X$. Comparing coefficients of $X^0 = 1$ on both sides of this equality, we have $b_0 = 2s_0$, i.e. b_0 must be an *even* integer. But since $Sb = J$ contains the polynomial X, we must have $X = gb = gb_0$, for some $g = g_0 + g_1 X + \ldots \in S$. Comparing coefficents of X, we get $1 = g_1 b_0$. This is impossible, because g_1, b_0 are integers, with b_0 divisible by 2. It follows that J is not a principal ideal.

8.11 The Division Theorem for polynomials

Let S be a commutative ring, and $S[X]$ the ring of all polynomials in X with coefficients in S. A typical element of $S[X]$ is a polynomial

$$f(X) = f_n X^n + f_{n-1} X^{n-1} + \ldots + f_1 X + f_0, \tag{8.14}$$

with coefficients $f_0, f_1, \ldots, f_n$ in S. In this we are departing from the notation used in expression (8.2) in two ways: first we write $f(X)$ in place of f, to emphasize that $f(X)$ is a polynomial in the 'variable' X, and secondly we arrange the terms $f_r X^r$ so that the highest power of X comes at the left.

A polynomial $g(X)$ of degree m is called *monic* if the coefficient $g_m = 1$, so that $g(X)$ has the form

$$g(X) = X^m + g_{m-1}X^{m-1} + \ldots + g_1X + g_0. \tag{8.15}$$

Notice that a monic polynomial is never zero, and that $1(= 1 + 0X + 0X^2 + \ldots)$ is the only monic polynomial of degree 0.

POLYNOMIAL DIVISION THEOREM. *Let $f(X)$, $g(X)$ be elements of $S[X]$, with $g(X)$ monic. Then there exist polynomials $q(X)$, $r(X)$ such that $f(X) = q(X)g(X) + r(X)$, where either $r(X)$ is zero, or deg $r(X) < m$.*

Remarks

1 $q(X)$ and $r(X)$ are called the *quotient* and *remainder*, respectively, when $f(X)$ is divided by $g(X)$. They correspond exactly to the quotient q and remainder r, which one gets by dividing an integer f by a positive integer g.
2 We should remember that the degree of a polynomial $r(X)$ is defined only when $r(X) \neq 0$.

Proof of the theorem. Let $f(X)$, $g(X)$ be as given in equations (8.14) and (8.15). If $f(X) = 0$, or if it has degree $n < m$, then the theorem holds, with $q(X) = 0$, $r(X) = f(X)$. Assume now that $f(X)$ has degree $n \geq m$. Then the polynomial $f_n X^{n-m} g(X)$ has the same leading term as $f(X)$, hence $f^{(1)}(X) = f(X) - f_n X^{n-m} g(X)$ is either zero, or else has degree $n(1) < n$. If $f^{(1)}(X) = 0$, or if $n(1) < m$, then the theorem holds with $q(X) = f_n X^{n-m}$ and $r(X) = f^{(1)}(X)$. If $n(1) \geq m$, we calculate a new polynomial $f^{(2)}(X) = f^{(1)}(X) - f_{n(1)}^{(1)} X^{n(1)-m} g(X)$. This is either zero, or else has degree $n(2) < n(1)$. If $n(2) \geq m$, we calculate $f^{(3)}(X) = f^{(2)}(X) - f_{n(2)}^{(2)} X^{n(2)-m} g(X)$, and so on until for some k we find that $f^{(k)}(X) = f^{(k-1)}(X) - f_{n(k-1)}^{(k-1)} X^{n(k-1)-m} g(X)$ is zero, or has degree $n(k) < m$. Now we can check that $f(X) - q(X)g(X) = f^{(k)}(X)$, where $q(X) = f_n X^{n-m} + f_{n(1)}^{(1)} X^{n(1)-m} + \ldots + f_{n(k-1)}^{(k-1)} X^{n(k-1)-m}$. So the theorem holds, with $r(X) = f^{(k)}(X)$. Notice that the degrees of $f(X)$, $f^{(1)}(X)$, $f^{(3)}(X)$, $\ldots$ are strictly descending, i.e. $n > n(1) > n(2) > \ldots$, and this guarantees that we must eventually reach an $f^{(k)}(X)$ which is either zero, or has degree $n(k) < m$.

Remark

3 From this proof, we see that if $\deg f(X) \geq m$, then $\deg q(X) = n - m$.

Example 206. Find the quotient and remainder when $f(X) = 3X^4 + 2X + 4$ is divided by $g(X) = X^2 - X + 1$. We set out the calculation just like a 'long division'. After $f(X)$ and $g(X)$ have

$$3X^2 + 3X \tag{a}$$

$$X^2 - X + 1 \overline{\big)\, 3X^4 + 0X^3 + 0X^2 + 2X + 4} \tag{b}$$

$$\underline{3X^4 - 3X^3 + 3X^2} \tag{c}$$

$$3X^3 - 3X^2 + 2X + 4 \tag{d}$$

$$\underline{3X^3 - 3X^2 + 3X} \tag{e}$$

$$-X + 4 \tag{f}$$

been written down in line (b), write $f_n X^{n-m} (= 3X^2)$ above the leading term $3X^4$ of $f(X)$. Next put $f_n X^{n-m} g(X)$ below $f(X)$, in line (c). Subtract this from $f(X)$ to give $f^{(1)}(X)$, line (d). This has degree $n(1) = 3$. Put $f^{(1)}_{n(1)} X^{n(1)-m} = 3X$ in line (a), above the leading term in (d). Multiply this by $g(X)$ and put the result in line (e). Subtract to get $f^{(2)}(X)$, in line (f). Since this has degree $< m$ ($m = 2$), we are finished. The result is $f(X) = (3X^2 + 3X)g(X) + (-X + 4)$, i.e. $q(X) = 3X^2 + 3X$ and $r(X) = -X + 4$.

Substitution maps. Zeros of polynomials

Let $f(X) \in S[X]$ and $a \in S$. We denote by $f(a)$ the element of S obtained by substituting a for X in $f(X)$, namely

$$f(a) = f_n a^n + f_{n-1} a^{n-1} + \ldots + f_1 a + f_0. \tag{8.16}$$

Notice that if $f(X) = f_0 + 0X + 0X^2 + \ldots$ (which means that $f(X)$ is zero, or has degree zero) then $f(a) = f_0$. In particular, if $f(X)$ is the polynomial 1, then $f(a) = 1$, for all $a \in S$.

For any given $a \in S$, the map $\Phi_a : S[X] \to S$ which takes each $f(X) \in S[X]$ to $f(a)$, is called a *substitution map*; we can check easily that Φ_a is a *ring-homomorphism*. **RH1** holds, since $\Phi_a(1) = 1$ by a remark just made; **RH2**, **RH3** hold, since if $q(X)$ and $p(X)$ are the sum and product of given polynomials $f(X)$, $g(X)$, it is clear from definitions (b), (c) of section 8.5 that $q(a)=f(a)+g(a)$ and $p(a)=f(a)g(a)$, i.e. $\Phi_a(f+g)=\Phi_a(f)+\Phi_a(g)$ and $\Phi_a(fg)=\Phi_a(f)\Phi_a(g)$.

In the *theory of polynomial equations*, one is given a polynomial $f(X)\in S[X]$, and the problem is to find all elements $a\in S$ such that $f(a)=0$. Such an element a is called a *zero of $f(X)$*, or a *root of the equation* $f(X)=0$ (it is inaccurate, although regrettably common, to refer to a as a 'root of $f(X)$'). The Polynomial Division Theorem gives information about the zeros of polynomials, through the following theorem.

REMAINDER THEOREM. *Let $f(X)\in S[X]$ be a polynomial of degree $n\geq 1$, and let $a\in S$. Then* (i) $f(X)=q(X)(X-a)+f(a)$, *for some $q(X)\in S[X]$ of degree $n-1$, and* (ii) *a is a zero of $f(X)$ if and only if $X-a$ divides $f(X)$.*

Proof. (i) Apply the Division Theorem with $g(X)=X-a$. Then $m=1$, and so

$$f(X)=q(X)(X-a)+r(X), \tag{8.17}$$

where either $r(X)=0$, *or* $r(X)$ has degree zero, i.e. $r(X)=r_0+0X+0X^2+\ldots$ for some $r_0\in S$. Also the degree of $q(X)$ is $n-1$, by Remark 3 above. Now apply the substitution map Φ_a to equation (8.17). Because Φ_a is a homomorphism we get $f(a)=q(a).0+r_0$, i.e. $r_0=f(a)$. Therefore $f(X)=q(X)(X-a)+f(a)$.

(ii) If a is a zero of $f(X)$, then $f(X)=q(X)(X-a)$ by (i), which shows that $X-a$ divides $f(X)$. Conversely if $X-a$ divides $f(X)$, we have $f(X)=h(X)(X-a)$ for some $h(X)\in S[X]$; then applying Φ_a we have $f(a)=h(a)(a-a)=h(a).0=0$.

Example 207. Take a positive integer r and let $f^{(r)}(X)=X^r-a^r$. Since $f^{(r)}(X)=a^r-a^r=0$, we know from part (ii) of the Remainder Theorem that $X-a$ divides $f^{(r)}(X)$. We can also prove this by direct calculation: $X^r-a^r=(X-a)(X^{r-1}+aX^{r-2}+\ldots+a^{r-2}X+a^{r-1})$.

Example 208. Let $S=Z/6Z$, as in Example 165 (section 8.1). We shall denote the elements of S as $0, 1, \ldots, 5$, in place of the clumsier $E_0, E_1, \ldots, E_5$ – we must remember, of course, that all calculations in $Z/6Z$ are done 'mod 6'. Let $f(X)=X^2+3X+2$; this is an element of $S[X]$.

Substituting for X the six elements of S in turn, we get $f(0)=2$, $f(1)=0$, $f(2)=0$, $f(3)=2$, $f(4)=0$, $f(5)=0$. Therefore $f(X)$ has four zeros, viz. 1, 2, 4, 5.

Example 209. Suppose that S is a subring of a commutative ring T. Then $S[X]$ is a subring of $T[X]$, since every polynomial $f(X) \in S[X]$ can be regarded as a polynomial with coefficients in T. It is quite possible for $f(X)$ to have zeros in T, which are not in S. For example take $S = R$ (real numbers) and $T = C$ (complex numbers). Then $f(X) = X^2 + 1 \in R[X]$ has no zeros in R, but has zeros i, $-i$ in C. We have $f(X) = (X-i)(X+i)$, where $X-i$, $X+i$ are elements of $C[X]$. But we cannot 'factorize' $f(X)$ as a product of polynomials of degree 1 in $R[X]$.

Example 210. The complex field C has the remarkable property that is *algebraically closed*; this means that every polynomial $f(X) \in C[X]$ with degree $n \geq 1$ has at least one zero in C, or in other words, any equation $f_n X^n + \ldots + f_1 X + f_0 = 0$ with coefficients $f_0, \ldots, f_n \in C$, has at least one root $\alpha \in C$. This is known as the *Fundamental Theorem of Algebra*, and is quite hard to prove without using fairly sophisticated analysis of complex functions (see section 7.4 of P. M. Cohn, *Algebra*, vol 2, J. Wiley, London; or pp. 513–15 of M. Spivak, *Calculus*, 2nd edn., M. Spivak, Waltham, Mass.). But if we assume it to be true, we have from the Remainder Theorem that every polynomial $f(X) \in C[X]$ of degree $n \geq 1$, can be written $f(X) = (X-\alpha_1)q(X)$ for some $\alpha_1 \in C$, and $q(X) \in C[X]$ of degree $n-1$. If $n \geq 2$, we apply the same argument to $q(X)$; we have $q(X) = (X-\alpha_2)h(X)$ for some $\alpha_2 \in C$ (which might be equal to α_1) and $h(X) \in C[X]$ of degree $n-2$. Eventually we find that $f(X) = (X-\alpha_1)(X-\alpha_2) \ldots (X-\alpha_n)c$, where $\alpha_1, \alpha_2, \ldots, \alpha_n \in C$ and c is a non-zero element of C. Comparing coefficients of X^n, we see that $c = f_n$. So *every polynomial in $C[X]$ of degree $n \geq 1$ can be factorized as a product of polynomials of degree 1*.

8.12 Polynomials over a field

So far we have allowed the coefficients of a polynomial to come from any commutative ring S. In this section we shall consider polynomials with coefficients from a *field* F. With this restriction, we can prove two important theorems (i), (ii) below which would not hold over an arbitrary commutative ring S. For the proof of theorem (i), the important point is that a field F is an *integral domain* (Example 175, section 8.4).

THEOREM. (i) *Let F be a field and $f(X) \in F[X]$ a non-zero polynomial of degree $n \geq 1$. If $a_1, a_2, \ldots, a_s$ are distinct zeros of $f(X)$, then $f(X) = (X - a_1)(X - a_2) \ldots (X - a_s)h(X)$, for some polynomial $h(X) \in F[X]$ of degree $n - s$. Hence $s \leq n$, i.e. the number of distinct zeros of a polynomial $f(X) \in F[X]$ is not greater than the degree of $f(X)$.*

Proof. This is by induction on s. By theorem (ii), section 8.11, taking $a = a_1$, we have $f(X) = (X - a_1)q(X)$ for some $q(X) \in F[X]$, which has degree $n - 1$. Thus our present theorem holds for $a = 1$. Now assume $s \geq 2$, and that the theorem holds for $s - 1$. Substituting a_i for X in the equation $f(X) = (X - a_1)q(X)$, we get $0 = f(a_i) = (a_i - a_1)q(a_i)$, for $i = 2, \ldots, s$. But $a_i - a_1 \neq 0$ (since $a_1, \ldots, a_s$ are distinct), hence *because F is an integral domain*, we have $q(a_i) = 0, i = 2, \ldots, s$. Applying the case $s - 1$ of our theorem, we have $q(X) = (X - a_2) \ldots (X - a_s)h(X)$, for some $h(X) \in F[X]$ of degree $(n - 1) - (s - 1) = n - s$. Hence $f(X) = (X - a_1)q(X) = (X - a_1)(X - a_2) \ldots (X - a_s)h(X)$, which proves the theorem.

Remark

Example 208 shows that theorem (i) fails when F is replaced by $S = Z/6Z$, since it exhibits a polynomial $f(X)$ of degree 2, which has four zeros.

Example 211. Case $s = n$ of theorem (i) says: if $f(X) \in F[X]$ has degree $n \geq 1$, and if $a_1, a_2, \ldots, a_n$ are n distinct zeros of $f(X)$, then $f(X) = (X - a_1)(X - a_2) \ldots (X - a_n)c$ for some $c \in F[X]$ of degree zero, i.e. c is a non-zero element of F. In fact by comparing coefficients of X^n in the above equation we see that $c = f_n$. In particular if $f(X)$ is monic, we have $f(X) = (X - a_1)(X - a_2) \ldots (X - a_n)$. An interesting example is the polynomial $f(X) = X^n - 1 \in C[X]$. This has n distinct zeros $1, \beta, \beta^2, \ldots, \beta^{n-1}$, where $\beta = \cos(2\pi/n) + i \sin(2\pi/n)$ (see W. Ledermann, *Complex Numbers*, p. 35). It follows that

$$X^n - 1 = (X - 1)(X - \beta) \ldots (X - \beta^{n-1}).$$

Example 212. A kind of 'converse' to theorem (i) is: if $a_1, a_2, \ldots, a_n$ are any elements of F *not necessarily distinct*, then the only zeros of the polynomial $f(X) = (X - a_1)(X - a_2) \ldots (X - a_n)$ are $a_1, a_2, \ldots, a_n$. For if b is any zero of $f(X)$ we have $0 = f(b) = (b - b_1)(b - a_2) \ldots (b - a_n)$. Since F is an integral domain, not all of $b-a_1, \ldots, b-a_n$ can be non-zero (this is because, in an integral domain, any product of non-zero elements is non-zero). So $b = a_i$ for some a_i. Notice that this fails when F is replaced by an arbitrary commutative ring S. For in Example 208, $f(X) = (X-1)(X-2)$ has zeros which are not equal to either of 1 or 2.

Example 213. Let $f(X) = f_n X^n + \ldots + f_1 X + f_0$ be a polynomial of degree $n \geq 1$, with coefficients $f_n, \ldots, f_1, f_0 \in R$; since $R \subseteq C$, we may regard $f(X)$ as element of $C[X]$. Suppose that $f(\alpha) = 0$, for some $\alpha = a_0 + i a_1$ in C which is not real, i.e. such that $a_1 \neq 0$; notice that α and $\bar{\alpha}$ are distinct. Then $f(\bar{\alpha}) = f_n \bar{\alpha}^n + \ldots + f_1 \bar{\alpha} + f_0 = \bar{f_n} \bar{\alpha}^n + \ldots \bar{f_1} \bar{\alpha} + \bar{f_0}$ (since $f_r = \bar{f_r}$, all $r \geq 0$) $= \overline{f(\alpha)}$ (since complex conjugation gives an automorphism of the ring C, Example 189, section 8.6). So *both* α and $\bar{\alpha}$ are zeros of $f(X)$, and by theorem (i) $f(X) = (X-\alpha)(X-\bar{\alpha})h(X)$, for some $h(X) \in C[X]$ of degree $n-2$. But $(X-\alpha)(X-\bar{\alpha}) = X^2 - (\alpha + \bar{\alpha})X + \alpha\bar{\alpha} = X^2 - 2a_0 X + (a_0^2 + a_0^2)$ is a monic polynomial with real coefficients; let us call this $p(X)$. Since $f(X) = p(X)h(X)$, $h(X)$ can be found by dividing $f(X)$ by $p(X)$, both of which lie in $R[X]$; therefore $h(X)$ lies in $R[X]$. So we have a theorem about real polynomials: *if $F(X) \in R[X]$ has degree $n \geq 1$, and if $f(\alpha) = 0$ for a complex number $\alpha = a_0 + i a_1$ which is not real, then $f(X) = p(X)h(X)$, where $p(X) = X^2 - 2a_0 X + (a_0^2 + a_1^2)$, and $h(X) \in R[X]$ has degree $n-2$.* For example, if $f(i) = 0$, it follows that $f(X)$ is divisible by $X^2 + 1$.

Example 214. Suppose that $f(X)$, $g(X)$ are polynomials in $F[X]$, both of degree $n \geq 1$, and that $f(a) = g(a)$ for $n+1$ distinct values $a = a_1$, $a_2, \ldots, a_{n+1} \in F$. Then $f(X)$ must be equal to $g(X)$. For $h(X) = f(X) - g(X)$ is a polynomial of degree at most n, and $h(X)$ has $n+1$ distinct zeros $a_1, \ldots, a_{n+1}$. If deg $h(X) \geq 1$, this contradicts theorem (i). So in fact $h(X) = h_0 \in F$; however $h(a_1) = 0$, which shows that $h_0 = 0$, hence $h(X)$ is the zero polynomial.

THEOREM. (ii) *Let F be a field. Then every ideal J of $F[X]$ is principal; in fact if $J \neq \{0\}$, then $J = F[X]g(X)$, where $g(X) \in F[X]$ is monic.*

Proof. If $J = \{0\}$ then $J = F[X]0$, which shows that J is a principal ideal. Now assume $J \neq \{0\}$, and from the set $\mathcal{N}$ of all non-zero polynomials in J choose one, $h(X)$, of smallest degree m. The coefficient h_m of $h(X)$ is non-zero, hence (since F is a field) there exists $h_m^{-1} \in F$. We can regard h_m^{-1} as an element of $f[X]$ of degree 0. By condition I2 (section 8.7), $g(X) = h_m^{-1}h(X)$ belongs to J; notice that $g(X)$ is monic, and of degree m. Since $g(X) \in J$, then $f(X)g(X) \in J$

for all $f(X) \in F[X]$, hence $F[X]g(X) \subseteq J$. To prove the converse inclusion, take any $f(X) \in J$. By the Division Theorem (section 8.11) there exist $q(X)$, $r(X)$ such that $f(X) = q(X)g(X) + r(X)$, and if $r(X) \neq 0$, then deg $r(X) < m$. But $r(X) = f(X) - q(X)g(X) \in J$, since $f(X) \in J$ and $q(X)g(X) \in J$. So the set $\mathcal{N}$ of non-zero elements of J contains a member $r(X)$ with degree $< m$; this contradicts our definition of m. Hence we must have $r(X) = 0$. Thus $f(X) = q(X)g(X) \in F[X]g(X)$, for all $f(X) \in J$, i.e. $J \subseteq F[X]g(X)$. Taken with what we proved above, this gives $J = F[X]g(X)$, and the theorem is proved.

Remark

2 This theorem fails if F is replaced by an arbitrary commutative ring, or even by an arbitrary integral domain – Example 205 gives an ideal J of $Z[X]$, which is not principal.

8.13   Divisibility in Z and in $F[X]$

A fundamental property of the ring Z is that every integer n (≥ 1) can be expressed, in essentially only one way, as a product $n = p_1 p_2 \ldots p_r$, where $p_1, p_2, \ldots, p_r$ are prime numbers. In this section we shall generalize this to a theorem which applies not only to Z but also to the ring $F[X]$ of polynomials over a field F.

The common link between Z and $F[X]$ is that they are both principal ideal domains. For the moment, however, we shall work with an arbitrary integral domain S. If $a, b \in S$, we say that b *divides* a (or b is a *factor* of a, or a is *divisible by* b, or a is a *multiple of* b; all these expressions mean the same thing!) if there is some $s \in S$ such that $a = sb$. Write $b|a$ to denote that b divides a. Next define a relation $\sim$ on S by the rule: $a \sim b$ if and only if $a|b$ *and* $b|a$. Elements a, b of S are said to be *associated* if $a \sim b$. We shall prove that $\sim$ is an equivalence relation in the next lemma, which shows how divisibility can be expressed in terms of ideals of S.

LEMMA. *Let a, b be elements of an integral domain S. Then*

(i) $b|a \Leftrightarrow Sa \subseteq Sb$.

(ii) $a \sim b \Leftrightarrow Sa = Sb$.

(iii) $a \sim b \Leftrightarrow a = ub$ *for some* $u \in U(S)$.
(iv) $\sim$ *is an equivalence relation on* S.

Proof. (i) If $b|a$ there is some $s \in S$ such that $a = sb$; then $xa = xsb \in Sb$ for all $x \in S$, hence $Sa \subseteq Sb$. Conversely if $Sa \subseteq Sb$ then $a \in Sb$, because $a = 1a \in Sa$. But $a \in Sb \Rightarrow a = sb$ for some $s \in S$, hence $b|a$.

(ii) By (i), $a|b$ and $b|a \Leftrightarrow Sb \subseteq Sa$ and $Sa \subseteq Sb \Leftrightarrow Sa = Sb$, hence $a \sim b \Leftrightarrow Sa = Sb$.

(iii) $a \sim b \Rightarrow a|b$ and $b|a \Rightarrow$ (there exist u, $t \in S$ such that $a = ub$ and $b = ta) \Rightarrow a = ub = uta \Rightarrow (1 - ut)a = 0$. If $a \neq 0$, we must have $1 - ut = 0$, because S is an integral domain, hence $ut = 1$, which means that u has inverse $u^{-1} = t$ in S, i.e. $u \in U(S)$. If $a = 0$, then $b = ta = 0$, and we can take $u = 1$; in either case $a = ub$ for some $u \in U(S)$. Conversely if $a = ub$ for some $u \in U(S)$, then we have also $b = u^{-1}a$; hence $b|a$ and $a|b$, i.e. $a \sim b$.

(iv) Check that $\sim$ satisfies **E1**, **E2**, **E3** – the easiest way is to use (ii).

Remarks

1 Elements a, b of S which are associated have the same 'divisibility behaviour'. Thus if $a \sim b$ and c is any element of S, then $c|a$ if and only if $c|b$, because $c|a \Leftrightarrow Sa \subseteq Sc$ (by (i)) $\Leftrightarrow Sb \subseteq Sc$ (since $Sa = Sb$ by (ii)) $\Leftrightarrow c|b$. Also $a|c$ if and only if $b|c$, because $a|c \Leftrightarrow Sc \subseteq Sa \Leftrightarrow Sc \subseteq Sb \Leftrightarrow b|c$.

2 Taking $b = 1$ in (iii) gives: $a \sim 1 \Leftrightarrow a \in U(S)$; taking $b = 0$ gives: $a \sim 0 \Leftrightarrow a = 0$. So the $\sim$-classes (i.e. the equivalence classes for $\sim$) of 1 and 0 are $U(S)$ and $\{0\}$, respectively.

We shall need to know what the relation $\sim$ becomes, in the special cases $S = Z$ and $S = F[X]$ (F a field). By (iii) above, the important thing is to know the group $U(S)$ of invertible elements in each case. By Example 166 (section 8.1), $U(Z) = \{1, -1\}$. Hence *if* $a, b \in Z$, then $a \sim b \Leftrightarrow a = b$ or $-b$. The $\sim$-classes for Z are $\{0\}$ and $\{n, -n\}$, $n = 1, 2, 3, \ldots$. Now consider the case $S = F[X]$. If $f(X) \in F[X]$ is invertible, there is some $g(X) \in F[X]$ such that $f(X)g(X) = 1$. This shows that $f(X), g(X)$ are both non-zero, and deg

$f + \deg g = \deg 1 = 0$ (theorem (ii), section 8.5). So $\deg f$ and $\deg g$ must both be 0. So $f(X) = f_0$, where f_0 is a non-zero element of F. Conversely such a polynomial $f(X)$ is invertible, with inverse $f(X)^{-1} = f_0^{-1}$ (we are using the fact that F is a field). So $U(F[X]) = \{c \in F | c \neq 0\}$, regarding $c \in F$ as the polynomial $c + 0X + 0X^2 + \ldots$. Hence *if $f(X)$, $g(X) \in F[X]$, then $f(X) \sim g(X) \Leftrightarrow f(X) = cg(X)$ for some $c \in F$, $c \neq 0$*. Notice that each $\sim$-class in $F[X]$ (apart from the class $\{0\}$) contains exactly one *monic* polynomial, since if $f(X)$ is any non-zero element of $F[X]$, there is exactly one value of $c \in F$ such that $cf(X)$ is monic, namely $c = f_n^{-1}$, where $n = \deg f$.

Irreducible elements

Let p be a non-zero element of an integral domain S. Any equation $p = ab$, with $a, b \in S$, is called a *factorization of p*; the factorization is said to be *trivial* if either $a \in U(S)$ or $b \in U(S)$.

DEFINITION. A non-zero element p of S is *irreducible* in S if $p \notin U(S)$ and every factorization of p is trivial, i.e.

$$p = ab \Rightarrow a \in U(S) \quad \text{or} \quad b \in U(S), \qquad \text{for all } a, b \in S. \quad (8.18)$$

Condition (8.18) is equivalent to

$$a | p \Rightarrow a \sim 1 \quad \text{or} \quad a \sim p, \qquad \text{for all } a \in S. \quad (8.19)$$

For suppose (8.18) holds and that $a | p$. There must be some $b \in S$ such that $p = ab$. By (8.18), either $a \in U(S)$, in which case $a \sim 1$ (by (iii)), or $b \in U(S)$, in which case $p \sim a$ (by (iii)). Therefore (8.19) holds. Conversely if (8.19) holds and $p = ab$ is any factorization of p, then by (8.19) either $a \sim 1$, i.e. $a \in U(S)$, or $a \sim p$, i.e. $a = pu$ for some $u \in U(S)$. In this last case $p = pub$, giving $ub = 1$, so that $b = u^{-1} \in U(S)$. Thus (8.18) holds.

Remarks

3 It follows from (8.19) and Remark 1, that if $p, q \in S$ and $p \sim q$, and p is irreducible if and only if q is irreducible. So if one element of a $\sim$-class is irreducible, all elements in that class are irreducible.

4 Notice that *by definition*, elements of $U(S)$ are not irreducible,

although they do satisfy conditions (8.18) and (8.19). In particular, 1 is not irreducible.

Irreducible elements of Z

Let p be a non-zero element of Z. Then $p \in U(Z) \Leftrightarrow p \in \{1, -1\} \Leftrightarrow |p| = 1$. So p is irreducible if and only if $|p| \neq 1$, and $p = ab \Rightarrow$ either $|a| = 1$ or $|b| = 1$. This implies that either p is a prime number, or it is $-$ (prime number), i.e. the irreducible elements of Z are 2, -2, 3, -3, 5, -5, ...

Irreducible elements of $F[X]$

Let $p(X)$ be a non-zero polynomial in $F[X]$. Then $p(X) \in U(F[X]) \Leftrightarrow \deg p(X) = 0$. So $p(X)$ is irreducible if $\deg p(X) \neq 0$ (i.e. $\deg p(X) \geq 1$), and for any factorization $p(X) = a(X)b(X)$ with $a(X)$, $b(X) \in F[X]$, either $\deg a(X) = 0$ or $\deg b(X) = 0$. Notice that any polynomial $p(X)$ of degree 1 is irreducible.

Example 215. If $p(X)$ is a monic polynomial, and if $p(X) = a(X)b(X)$, with $a(X) = a_h X^h + \ldots + a_0$, $b(X) = b_k X^k + \ldots + b_0$, then by comparing highest terms in the equation $p(X) = a(X)b(X)$, we get $X^{h+k} = a_h X^h b_k X^k$, hence $1 = a_h b_k$. So if we define $a^*(X) = a_h^{-1} a(X)$ and $b^*(X) = b_k^{-1} b(X)$, we find $p(X) = a^*(X)b^*(X)$, and $a^*(X)$, $b^*(X)$ are both *monic*. Since $a^*(X)$, $b^*(X)$ have the same degrees as $a(X)$, $b(X)$, respectively, it follows that in checking that a *monic* $p(X)$ is irreducible by the condition above, we have only to consider factorizations $p(X) = a(X)b(X)$ where both $a(X)$, $b(X)$ are themselves monic.

Example 216. The only irreducible, monic polynomials $p(X)$ in $C[X]$ are those of degree 1, i.e. $p(X) = X + p_0$ ($p_0 \in C$). For suppose $p(X)$ has degree $n \geq 2$. By the Fundamental Theorem of Algebra (Example 210) $p(X)$ has a zero in C, say α. But then the Remainder Theorem gives $p(X) = (X - \alpha)q(X)$ for some $q(X) \in C[X]$ of degree $n - 1$. So $p(X)$ is not irreducible in $C[X]$.

Example 217. Let p be a prime number, so that $F_p = Z/pZ$ is a field with p elements E_0, E_1, ..., E_{p-1} (Example 177, section 8.4). For ease of calculation, we shall write these elements simply as 0, 1, ..., $p-1$ – with the warning that all calculations with these 'integers' must be made mod p. If p is small, it is often quite easy to decide whether a given polynomial in

$F_p[X]$ of small degree is irreducible or not. For example, take $f(X) = X^3 + X + 1 \in F_5[X]$. If $f(X)$ *is not* irreducible, then it has a factorization $f(X) = a(X)b(X)$, where both $a(X)$ and $b(X)$ have degrees ≥ 1; we can assume (Example 215) that $a(X)$, $b(X)$ are both monic. Since deg $a(X)$ + deg $b(X) = 3$, and since we may interchange $a(X)$ and $b(X)$ without altering their product, we may assume deg $a(X) = 1$, deg $b(X) = 2$; we have $f(X) = (X + a_0)(X^2 + b_1 X + b_0)$ for some $a_0, b_1, b_0 \in F_5$. But this shows that $f(-a_0) = 0$. By direct calculation we find $f(0) = 1$, $f(1) = 3$, $f(2) = 1$, $f(3) = 1$, $f(4) = 4$ (all calculations mod 5). So there is no element $a_0 \in F_5$ such that $f(-a_0) = 0$. Hence $f(X)$ is irreducible in $F_5[X]$.

Highest common factors

If a, b are non-zero elements of an integral domain S, then any element $d \in S$ is called a *highest common factor* (hcf) of a and b if it satisfies the two conditions

HCF1 $d|a$ and $d|b$,

and

HCF2   If $e \in S$ satisfies $e|a$ and $e|b$, then $e|d$.

It is by no means obvious that a given pair (a, b) of elements of S have an hcf. But if they do, then this is unique in the following sense: *If d, d' are both hcf's of a and b, then $d \sim d'$.* For if we take $e = d'$ in **HCF2**, we get $d'|d$. But since our assumptions on d and d' are identical, we must also have $d|d'$. So $d|d'$ and $d'|d$, i.e. $d \sim d'$.

THEOREM. (i) *Let a, b be non-zero elements of a PID S. Then there exists a hcf d of a and b. For any hcf d' of a and b, there exist $x, y \in S$ such that $xa + yb = d'$.*

Proof. By the lemma in section 8.7, the sum $Sa + Sb$ of the principal ideals Sa and Sb, is itself an ideal of S. Since every ideal in S is principal, there must exist $d \in S$ such that $Sd = Sa + Sb$. Since $Sd \supseteq Sa$ and $Sd \supseteq Sb$, we have $d|a$ and $d|b$, i.e. **HCF1** holds for d. Now suppose $e \in S$ is such that $e|a$ and $e|b$. Then $Se \supseteq Sa$ and $Se \supseteq Sb$, so $Se \supseteq Sa + Sb = Sd$, hence $d|e$. Thus **HCF2** holds as well; this shows that any element d of S such that $Sd = Sa + Sb$ is a hcf of a and b. If d' is any hcf of a and b, then we have $d \sim d'$ (see above). So

$Sd' = Sd = Sa + Sb$. Thus $d' = 1d' \in Sa + Sb$, so there must be elements $x, y \in S$ such that $d' = xa + yb$.

Example 218. Taking $S = Z$, we see that 6 is an hcf of 18 and 30, by checking **HCF1**, **HCF2**. We can find $x, y \in Z$ such that $18x + 30y = 6$, e.g. $x = 2, y = -1$.

In the next section we shall give an explicit construction or 'algorithm' to find the hcf of given elements a, b in the cases $S = Z$ and $S = F[X]$.

THEOREM. (ii) *Let S be a PID and a, b be non-zero elements of S. Let* $p \in S$ *be irreducible. If* $p|ab$, *then* $p|a$ *or* $p|b$, *or both.*

Proof. Suppose that $p|ab$, but that p does not divide a. Let d be a hcf of p and a – this exists by theorem (i). Since $d|p$ and p is irreducible, we have $d \sim 1$ or $d \sim p$. But $d|a$, so we cannot have $d \sim p$, since this would imply $p|a$ (see Remark 1, p. 149), contrary to our assumption. So $d \sim 1$, which means that 1 is an hcf of a and p. By theorem (i) there exist elements $x, y \in S$ such that $1 = xa + yp$, hence $b = xab + ypb$. But since p divides ab, it follows that p divides b. This proves the theorem, for we have shown that if p does not divide a, then it divides b.

Corollary to theorem (ii). If $p|a_1 a_2 \ldots a_r$, where $a_1, a_2, \ldots, a_r$ are non-zero elements of S, then $p|a_i$ for at least one $i \in \{1, 2, \ldots, r\}$.

Proof. If $p \nmid a_1$ then $p|a_2 \ldots a_r$ (taking $a = a_1$, $b = a_2 \ldots, a_r$ in the theorem). If $p \nmid a_2$ then $p|a_3 \ldots a_r$, and so on. We must reach some $i \in \{1, 2, \ldots, r\}$ for which $p|a_i$.

Unique factorization theorem for Z and F[X]

For the rest of this section *we assume that S is either Z or F[X]* (*F* any field). Suppose $x \in S$ is such that $x \neq 0$ and $x \in U(S)$. We shall prove that x has at least one *irreducible factorization* (sometimes called a *prime factorization*)

$$x = p_1 p_2 \ldots p_r, \tag{8.20}$$

where $p_1, p_2, \ldots, p_r$ are irreducible elements of S (not necessarily distinct). First take the case $S = Z$. Since $x \neq 0$ and $x \notin U(Z)$ we have $|x| \geq 2$. If $|x| = 2$, i.e. if $x = 2$ or -2, then x is itself irreducible, so (8.20) holds with $r = 1$ and $p_1 = x$. We use induction on $|x|$: assume that $|x| > 2$ and that any element $y \in Z$, such that $y \neq 0$, $y \notin U(Z)$ and $|y| < |x|$, has an irreducible factorization. If x is itself irreducible, we can find a factorization (8.20), as before. So assume x is not irreducible. This implies that there is a factorization $x = yz$ which is *non-trivial*, i.e. $y \notin U(Z)$, $z \notin U(Z)$. We also have $y \neq 0$, $z \neq 0$ since $S = Z$ is an integral domain. But then $|y|$, $|z|$ are both ≥ 2, hence $|y| < |x|$, $|z| < |x|$. Applying our induction assumption, we see that both y, z have irreducible factorizations, and it follows that $x = yz$ has an irreducible factorization.

The case $S = F[X]$ is proved in a similar way. Since $x \neq 0$ and $x \notin U(F[X])$, we have $\deg x \geq 1$. If $\deg x = 1$ then x is itself irreducible, so (8.20) holds with $r = 1$ and $p_1 = x$. We now use induction on the degree $\deg x$ of the polynomial x; the details are left as an exercise for the reader.

The next theorem shows that there is 'essentially' only one way of making an irreducible factorization (8.20) of x; it is often called the *Unique Factorization Theorem*.

THEOREM. (iii) *Let S be either Z or $F[X]$ (F any field) and let $x \in S$ be such that $x \neq 0$ and $x \notin U(S)$. If $x = p_1 p_2 \ldots p_s = q_1 q_2 \ldots q_r$, where the p_i and the q_j are all irreducible elements of S, then $s = r$, and, after suitably rearranging the q_j's, we have $p_i \sim q_i$, for $i = 1, \ldots r$.*

Proof. This is by induction on s. If $s = 1$, then $x = p_1$ is irreducible, and so has no non-trivial factorization. This shows that $r = 1$; hence $p_1 = q_1$, and the theorem holds. Now assume that $s \geq 2$, and that the theorem holds in any case where the number of p_i's is $s - 1$. Since p_1 divides $q_1 q_2 \ldots q_r$, the corollary above shows that $p_1 | q_i$, for some $i \in \{1, 2, \ldots, r\}$. Since q_i is irreducible and $p_1 \not\sim 1$, we have $p_1 \sim q_i$. Now rearrange the q_j's so that q_i becomes the new q_1; we may now say that $p_1 \sim q_1$, hence there is some $u \in U(S)$ such that $q_1 = up_1$. From the equation $p_1 p_2 \ldots p_s = (up_1)q_2 \ldots q_r$ we may cancel p_1 (theorem, section 8.3) and get $p_2 \ldots p_s = (uq_2) \ldots q_r$.

Since the term on the left has $s-1$ irreducible factors, we may apply the theorem to this case, and deduce that $s-1=r-1$, and that after suitable rearranging the q_j's we have $p_2 \sim uq_2$, $p_3 \sim q_3, \ldots, p_r \sim q_r$. But since $uq_2 \sim q_2$, this shows that the theorem holds.

Example 219. Suppose we have $p_1(X) \ldots p_r(X) = q_1(X) \ldots q_s(X)$, where the $p_i(X)$ and $q_j(X)$ are all irreducible, *monic* polynomials in $F[X]$. Since $F[X]$ is a PID, we may apply the last theorem, which shows that $r=s$ and, after suitably rearranging the $q_j(X)$'s, $p_i(X) \sim q_i(X)$ for $i=1, \ldots, r$. However, since two monic polynomials are associated only if they are equal (see p. 150), we can say in this case that $p_i(X) = q_i(X)$ for $i=1, 2, \ldots, r$. A similar situation holds for factorizations in Z, if we assume that the p_i and q_j are all *positive* irreducible integers, i.e. that they are ordinary prime numbers.

8.14 Euclid's algorithm

Theorem (i) of the last section (p. 152) says that any non-zero elements a, b of a principal ideal domain S have a highest common factor $d \in S$, and that there exist elements x, $y \in S$ such that

$$xa + yb = d. \qquad (8.21)$$

Equation (8.21) has many interesting applications (see the examples below), and it is desirable to have a direct method of calculating x, y, d when a, b are given. We shall give such a method, called *Euclid's algorithm*, which works in the two important cases (I) $S=Z$, and (II) $S=F[X]$ (F a field). The theory in this section does not depend on the notion of an ideal.

Euclid's algorithm works by repeated use of the Division Theorem, which we formulate in such a way that it can be applied in either case (I) or (II).

DIVISION THEOREM. *Let S be either Z or $F[X]$ (F any field), and let f, g be non-zero elements of S. If $S=Z$, assume that $g>0$. Then there exist elements q, $r \in S$ such that*

$$f = qg + r, \qquad (8.22)$$

and r satisfies (I) $0 \leq r < g$ if $S = Z$, or (II) either $r = 0$ or $\deg r < \deg g$, if $S = F[X]$.

In both cases (I), (II) one refers to q as the 'quotient' and r as the 'remainder' when f is divided by g.

Proof. Case (I) $S = Z$. This is well known from elementary arithmetic. To give a formal proof, define P to be the set of all *non-negative* integers of the form $f - zg$, where $z \in Z$. P is not empty, since we can make $f - zg$ non-negative by taking z to be a sufficiently large negative integer. So there is some smallest element of P; call this r, so that $r = f - qg$ for some $q \in Z$. Then $r \geq 0$ because $r \in P$, and $r < g$ since otherwise we would have $r \geq g$, hence $r - g = f - (q + 1)g$ would be an element of P which is smaller than r, and this is impossible.

Case (II) $S = F[X]$

Our theorem is now almost the same as the Polynomial Division Theorem (p. 142); the difference is that we do not assume (as we did for the Polynomial Division Theorem) that $g = g(X) = g_m X^m + \cdots + g_1 X + g_0$ is *monic*. On the other hand we assume that the coefficients of $g(X)$ are in the *field* F, whereas for the Polynomial Division Theorem the coefficients came from an arbitrary commutative ring. So if $g(X)$ has degree m (which implies that $g_m \neq 0$) we know that g_m has inverse $g_m^{-1} \in F$. To find polynomials q, r satisfying (8.22), simply go through the proof on p. 142, replacing $g(X)$ by $g_m^{-1} g(X)$ wherever it occurs. The argument on those pages will still be valid, because the polynomial $g_m^{-1} g(X) = X^m + g_m^{-1} g_{m-1} X^{m-1} + \cdots + g_m^{-1} g_1 X + g_m^{-1} g_0$ is monic. We find $f(X) = q(X) g_m^{-1} g(X) = r(X)$ (in the notation of p. 142), hence (8.22) holds, with $q = q(X) g_m^{-1}$ and $r = r(X)$.

Euclid's algorithm

Let S be either Z or $F[X]$ (F any field), and let a, b be non-zero elements of S. If $S = Z$, assume that $b > 0$. Now define a sequence of elements $r_0, r_1, r_2, \ldots$ of S by the rule: let $r_0 = a$, $r_1 = b$, and when

r_i, r_{i+1} are given and $r_{i+1} \neq 0$, define r_{i+2} to be the remainder when r_i is divided by r_{i+1}. We have then a sequence of equations

$$a = q_1 b + r_2, \tag{8.23.0}$$

$$b = q_2 r_2 + r_3, \tag{8.23.1}$$

$$r_2 = q_3 r_3 + r_4, \tag{8.23.2}$$

$$\cdot \quad \cdot \quad \cdot$$
$$\cdot \quad \cdot \quad \cdot$$
$$\cdot \quad \cdot \quad \cdot$$

$$r_{k-3} = q_{k-2} r_{k-2} + r_{k-1}, \tag{8.23.k-3}$$

$$r_{k-2} = q_{k-1} r_{k-1} + r_k, \tag{8.23.k-2}$$

$$r_{k-1} = q_k r_k. \tag{8.23.k-1}$$

Here equation $(8.23.i)$ is the result of dividing r_i by r_{i+1}; the quotient is q_{i+1} and the remainder r_{i+2}. We can make this equation *provided r_{i+1} is not zero*, and then either $r_{i+1} < r_i$ in case (I) ($S = Z$), or $\deg r_{i+1} < \deg r_i$ in case (II) ($S = F[X]$). So the algorithm gives us a strictly descending sequence of non-negative integers in each case: in case (I) $S = Z$, we have $r_1 = b > r_2 > r_3 > \ldots$, and in case (II) $S = F[X]$, $\deg r_1 (= \deg b) > \deg r_2 > \deg r_3 > \ldots$. This sequence must end in a finite number of steps, which means there is some integer $k \geq 1$ such that $r_k \neq 0$, $r_{k+1} = 0$. The theorem below will show that this last non-zero remainder r_k is a highest common factor of a and b. At the same time, we shall see how to calculate, from the algorithm, elements $x, y \in S$ such that $xa + yb = r_k$.

THEOREM. *Let S be either Z or $F[X]$ (F any field) and let a, b be non-zero elements of S. If $S = Z$, assume $b > 0$. Let r_k be the last non-zero term in the sequence $r_0, r_1, r_2, \ldots$ defined by Euclid's algorithm. Then $r_k = d$ is a highest common factor of a, b. Moreover for each $i = k-1$, $k-2, \ldots, 1$ there are elements $X_i, Y_i \in S$ such that $X_i r_{i-1} + Y_i r_i = r_k$. In particular $xa + yb = d$, where $x = X_1$, $y = Y_1$.*

Proof. Let $d = r_k$. From $(8.23.k-1)$ we have $d | r_{k-1}$. Since d divides both r_{k-1} and r_k, we have $d|r_{k-2}$ from (8.23. $k-2$). Since d divides r_{k-2} and r_{k-1}, we have $d|r_{k-3}$ from (8.23.$k-3$). Going on like this we get finally that d divides $r_1 = b$ and $r_0 = a$. So d satisfies the condition **HCF1** (p. 152). We shall postpone the proof that d satisfies **HCF2** until we have proved the existence of elements X_i, Y_i as described in the theorem. For $i = k-1$ we do have elements X_i, Y_i satisfying $X_i r_{i-1} + Y_i r_i = r_k$: namely we take $X_{k-1} = 1$, $Y_{k-1} = -q_{k-1}$ and use (8.23.$k-2$). And if we have found X_i, Y_i for any i in the range $2 \leq i \leq k-1$, we can find suitable X_{i-1}, Y_{i-1} as follows:

$$r_k = X_i r_{i-1} + Y_i r_i$$
$$= X_i r_{i-1} + Y_i (r_{i-2} - q_{i-1} r_{i-1}) \text{ (using (8.23.}i-2))$$
$$= X_{i-1} r_{i-2} + Y_{i-1} r_{i-1},$$

where $X_{i-1} = Y_i$ and $Y_{i-1} = X_i - q_{i-1} Y_i$.

This proves that elements X_i exist for $i = k-1, k-2, \ldots, 1$. In particular $xa + yb = d$, where $x = X_1$, $y = Y_1$. It is now clear that d satisfies **HCF2**, since if $e \in S$ satisfies $e|a$ and $e|b$, then from $d = xa + yb$ it follows that $e|d$. The proof of the theorem is complete.

Remark

There are many pairs (x', y') of elements of S which satisfy $x'a + y'b = d$. For example, if x, y are the elements found above, we could take $x' = x + zb$, $y' = y - za$, for any $z \in S$. For applications, the important thing is to be able to find *some* pair (x, y) satisfying $xa + yb = d$.

Example 220. $(S = Z)$ Find an hcf d of $a = 540$, $b = 168$; find also integers x, y such that $xa + yb = d$. The Euclid algorithm gives $540 = 3.168 + 36$, $168 = 4.36 + 24$, $36 = 1.24 + 12$, $24 = 2.12$. The last non-zero remainder is $d = 12$, so this is an hcf of 540 and 168. Now working back through these equations (starting with $36 = 1.24 + 12$) we get $12 = 36 - 1.24 = 36 - 1(168 - 4.36) = -1.168 + 5.36 = -1.168 + 5(540 - 3.168) = 5.540 - 16.168$, i.e. $5a - 16b = d$.

Example 221. $(S = R[X])$ Find a hcf of $a(X) = X^3 - X$, $b(X) = 2X^2 - X - 1$.

The algorithm gives

$$a(X) = (\tfrac{1}{2}X + \tfrac{1}{4})b(X) + (-\tfrac{1}{4}X + \tfrac{1}{4}),$$

$$b(X) = (-8X - 4)(-\tfrac{1}{4}X + \tfrac{1}{4}) + 0,$$

hence $d(X) = r_2(X) = -\tfrac{1}{4}X + \tfrac{1}{4}$ is a hcf of $a(X)$ and $b(X)$. For greater elegance, we could take $d^*(X) = (-4)d(X) = X - 1$; this is also a hcf of $a(X)$, $b(X)$, because $d^*(X) \sim d(X)$ (see p. 152). The first equation gives polynomials x, y such that $xa + yb = d$, for it shows that $d(X) = 1.a(X) - (\tfrac{1}{2}X + \tfrac{1}{4})b(X)$. Multiplying by -4, we get $X - 1 = (-4).a(X) + (2X + 1).b(X)$.

DEFINITION. Non-zero elements a, b of S are said to be *co-prime*, or *relatively prime*, if 1 is a highest common factor of a, b.

From the Euclid algorithm (assuming $S = Z$ or $S = F[X]$), if a, b are co-prime there exist x, $y \in S$ such that

$$xa + yb = 1. \tag{8.24}$$

Conversely if a, b are such that (8.24) holds, for some x, $y \in S$, then a, b are co-prime. For any hcf d of a, b must divide 1 by equation (8.24), hence d is invertible. Therefore $d \sim 1$ (see Remark 2, p. 149), and it follows that 1 is a hcf of a, b, i.e. a, b are co-prime.

Example 222. If a, b are non-zero elements of S with hcf d, and if $a = da'$, $b = db'$ (a', $b' \in S$), then a', b' are co-prime. For we know there are x, $y \in S$ such that $ax + by = d$, i.e. $d(a'x + b'y) = d.1$. Since S is an integral domain we may cancel d and get $a'x + b'y = 1$, proving that a', b' are co-prime by the argument just given.

Example 223. Invertible elements of Z/mZ. Let m be a positive integer. The quotient ring Z/mZ has m elements $[0]$, $[1]$, ..., $[m-1]$, where $[a] = a + mZ = E_a$ (see Example 197, p. 136). The condition for $[a]$ to be invertible in Z/mZ is that there should exist some $[x]$ in Z/mZ with $[a][x] = 1$, i.e. that there should be some $x \in Z$ which satisfies the 'congruence equation'

$$ax \equiv 1 \bmod m. \tag{8.25}$$

We shall prove that $[a]$ is invertible, i.e. that (8.25) has a solution $x \in Z$, if and only if a is co-prime to m. First suppose (8.25) has a solution x, then

$ax - 1 \in mZ$, i.e. there is some $y \in Z$ such that $ax + my = 1$. This implies that a, m are co-prime. Conversely if a, m are co-prime then there exist x, $y \in Z$ such that $ax + my = 1$, hence $ax \equiv 1 \bmod m$ and so (8.25) has a solution. Notice that Euclid's algorithm gives an explicit way of calculating x and y.

We have now proved that $U(Z/mZ) = \{[a] | a \text{ is co-prime to } m\}$. For example if $m = 6$ (see Example 165, p. 113) the group of invertible elements of Z/mZ is $\{[1], [5]\}$. If $m = p$ is a prime, we get $U(Z/pZ) = \{[1], [2], \ldots, [p-1]\}$, i.e. every non-zero element of Z/pZ is invertible, so that Z/pZ is a field (we already proved this in Example 177, section 8.4).

Example 224. Find the inverse of [8] in the field $Z/43Z$. We must find $x \in Z$ such that $8x \equiv 1 \bmod 43$. We know that 8 and 43 are co-prime (because 43 is prime), but we use the Euclid algorithm to get x, y such that $8x + 43y = 1$. The algorithm gives $43 = 5.8 + 3$, $8 = 2.3 + 2$, $3 = 1.2 + 1$, $2 = 2.1$. As expected, this gives $d = 1$. Working back with these equations we get:
$1 = 3 - 1.2 = 3 - 1(8 - 2.3) = -1.8 + 3.3 = 1.8 + 3(43 - 5.8) = 3.43 - 16.8$.
So $-16.8 = 1 - 3.43 \equiv 1 \bmod 43$, hence $[8]^{-1} = [-16] = [27]$.

Example 225. Let n be a positive integer and $G = \text{gp}\{g\}$ be a cyclic group of order n. According to the discussion in section 5.4, every element of G is a power of g, and if a, $b \in Z$ we have $g^a = g^b$ if and only if $a \equiv b \bmod n$. In particular, $g^a = e$ if and only if $a \equiv 0 \bmod n$ (note: e is the unit element of G). We shall prove that, for any $a \in Z$, the subgroup $\text{gp}\{g^a\}$ of G which is generated by g^a, is also generated by g^d where d is the hcf of a and n (i.e. d is the *positive* hcf of a and n). Let $H = \text{gp}\{g^a\}$. We have x, $y \in Z$ such that $d = xa + yn$. So $g^d = g^{xa} g^{yn} = (g^a)^x . (g^n)^y = (g^a)^x . e^y = (g^a)^x$, hence $g^d \in H$. It follows that $\text{gp}\{g^d\} \subseteq H$. But since $d | a$ we have $a = da'$ for some $a' \in Z$. So $g^a = (g^d)^{a'} \in \text{gp}\{g^d\}$, hence $H \subseteq \text{gp}\{g^d\}$. Thus $H = \text{gp}\{g^d\}$, which is what we wanted to prove.

It is clear that the order of g^d is the integer n' such that $n = dn'$, since for any integer m the condition $(g^d)^m = e \Leftrightarrow dm \equiv 0 \bmod n \Leftrightarrow n$ divides $dm \Leftrightarrow n'$ divides $m \Leftrightarrow m \equiv 0 \bmod n'$. But since g^a generates the same subgroup H as g^d, then g^a has the same order as g^d. From these we deduce the following theorem: *if an element g of a group G has finite order n, then for any $a \in Z$, the order of g^a is n/d, where d is the hcf of a and n.*

Example 226. Construction of fields. Suppose that F is any field, and that $g(X) = X^m + g_{m-1}X^{m-1} + \cdots + g_1 X + g_0$ is an irreducible polynomial in $F[X]$. Let J denote the principal ideal $g(X)F[X]$. *Then the quotient ring*

$F[X]/J$ *is a field.* For let $[a(X)]$ denote the coset $a(X)+J$, for any $a(X) \in F[X]$. If $[a(X)]$ is not the zero element of $F[X]/J$, then $a(X)$ is not an element of J, i.e. $a(X)$ is not a multiple of $g(X)$. It follows that $a(X)$ and $g(X)$ are co-prime, because every divisor $d(X)$ of $g(X)$ must satisfy $d(X) \sim 1$ or $d(X) \sim g(X)$, and if $d(X)$ is also a divisor of $a(X)$ we cannot have $d(X) \sim g(X)$. So 1 is a hcf of $a(X)$ and $g(X)$, hence there exist $u(X)$, $v(X) \in F[X]$ such that $u(X)a(X)+v(X)g(X)=1$ – and we could find such polynomials $u(X)$ and $v(X)$ by applying the Euclid algorithm to $a(X)$ and $g(X)$. But then $u(X)a(X)-1=-v(X)g(X) \in J$, so that $[u(X)][a(X)]=1$, which shows that $[a(X)]$ has an inverse $[u(X)]$ in $F[X]/J$. Therefore $F[X]/J$ satisfies condition **F2** (section 8.4). It satisfies **F1**, since if we have $[1]=[0]$ it would imply that $1=1-0 \in J$ (see rule 8.13)), i.e. that $g(X)$ divides 1. This is impossible since $g(X)$ has degree $m \geq 1$. Finally $F[X]/J$ satisfies **F3**, because like $F[X]$ it is commutative. Therefore $F[X]/J$ is a field.

This construction allows us to make a large number of fields. In Example 200 (p. 137), we have $F=R$, the real field, and the irreducible polynomial is $g(X)=X^2+1$. In that case we found that the field $R[X]/J$ is isomorphic to the field C of complex numbers.

Exercises for Chapter 8

1 Write out the addition and multiplication tables for the ring $Z/5Z$. Show that the group $U(Z/5Z)$ of invertible elements of $Z/5Z$ is cyclic.

2 Let X be any set, and let $\mathscr{B}(X)$ be the set of all subsets of X. Define operations of addition and multiplication on $\mathscr{B}(X)$ by the rules: $A+B=A \oplus B=(A \cup B)-(A \cap B)$, and $AB=A \cap B$, for any subsets A, B of X. Prove that $\mathscr{B}(X)$, with these operations, is a commutative ring. [See Exercise 7, p. 14.] ($\mathscr{B}(X)$ is called a *Boolean ring.*)

3 Let X be any set and let S be the set of all maps $f:X \to R$ (R is the ring of real numbers). Define addition and multiplication on S by the rules: if f, $g \in S$, let $f+g$, fg be the elements of S given by $(f+g)(x)=f(x)+g(x)$, $(fg)(x)=f(x)g(x)$, for all $x \in X$. Prove that S is a commutative ring.

4 The *centre* $Z(S)$ of a ring S is defined to be the set of all $z \in S$ such that $zs=sz$ for all $s \in S$. Prove that $Z(S)$ is a subring of S.

5 An element e of a ring S is called an *idempotent* if $e^2=e$. Find all

the idempotents in the ring $Z/6Z$ (see Example 165, p. 113). Prove that if e is an idempotent in any commutative ring S, then the set $U = eS = \{es | s \in S\}$ satisfies conditions **SR2, SR3** but does not satisfy **SR1** in general. Show that eS is a ring, with e as identity element.

6   Show that $\text{rg}\{\sqrt{2}\}$, the subring of R generated by $\sqrt{2}$ (see Example 172, p. 117) consists of all real numbers of the form $z_0 + z_1\sqrt{2}$ $(z_0, z_1 \in Z)$.

7   Prove that any subring U of a field F is an integral domain.

8   If U is any subring of a field F, define $\hat{U}$ to be the set of all elements $s \in F$ which can be written in the form $s = uv^{-1}$, with $u, v \in U$ and $v \neq 0$. Prove that $\hat{U}$ is a subfield of F. [First prove that $\hat{U}$ is a subring.]

9   Let $U = \text{rg}\{\sqrt{2}\}$ be the subring of R given in Exercise 6. Show that $\hat{U}$ (see preceding exercise) consists of all real numbers of the form $r_0 + r_1\sqrt{2}$, with r_0, r_1 rational numbers.

10   Let $\alpha = a_0 + ia_1 + ja_2 + ka_3$ be any quaternion (see Example 179, p. 123). Verify that $\alpha\bar{\alpha} = N(\alpha)$, where $N(\alpha) = a_0^2 + a_1^2 + a_2^2 + a_3^2$. Prove that α satisfies the equation $\alpha^2 - 2a_0\alpha + N(\alpha) = 0$.

11   Prove that $N(\alpha)N(\beta) = N(\alpha\beta)$, for any quaternions α, β.

12   (Practice in working with complex numbers.) Let $u = 2 + i$, $v = 1 - i$. Calculate in turn $u + v$, uv, $(uv)^{-1}$, $u + v/uv$ (the last to be calculated as $(u + v)(uv)^{-1}$). Calculate also u^{-1}, v^{-1}; check that $u^{-1} + v^{-1} = u + v/uv$.

13   (S is any commutative ring.) Verify that the set $S[X]$ of all polynomials in X with coefficients in S, satisfies the ring axiom **R7**, i.e. that $(f + g)h = fh + gh$, for all $f, g, h \in S[X]$.

14   (S is any commutative ring.) Find a polynomial $f \in S[X]$ such that $f^2 = 1 - 2X + 5X^2 - 4X^3 + 4X^4$.

15   Assume that S is an integral domain. Show that if $f \in S[X]$ satisfies $f^2 = f$, then either $f = 1$ or $f = 0$.

16   Let S, T be any rings and $\theta: S \to T$ a ring homomorphism. Prove that if $x \in U(S)$, then $\theta(x) \in U(T)$, and $\theta(x^{-1}) = \theta(x)^{-1}$.

17   Let Q be the field of all rational numbers. Prove that the only ring-endomorphism $\theta: Q \to Q$ is the identity map on Q.

18   Define the map $\theta: H \to H$ (H is the quaternion ring, Example 179, p. 123) by $\theta(\alpha) = \bar{\alpha}$. Is θ a ring automorphism of H?

19 Prove that there is no ring homomorphism $\theta: R \to Z$.

20 Let S, T be commutative rings, and let $\theta: S \to T$ be a ring homomorphism. Prove that the map $\phi: S[X] \to T[X]$ given by the rule $\phi(f_0 + f_1 X + \cdots + f_n X^n) = \theta(f_0) + \theta(f_1)X + \cdots + \theta(f_n)X^n$, for any $f \in [X]$, is a ring homomorphism.

21 Prove that a simple, commutative ring S must be a field.

22 Show that if J is an ideal of S, and if $\mathcal{K}$ is an ideal of the quotient ring S/J, then the set $K = \{a \in S \mid a + J \in \mathcal{K}\}$ is an ideal of S. Prove also that $J \subseteq K \subseteq S$.

23 An ideal J of a ring S is said to be *maximal in S* if (i) $J \neq S$ and (ii) there are no ideals K of S satisfying $J \subseteq K \subseteq S$, except $K = J$ and $K = S$. Find all the maximal ideals in Z.

24 Show that if J is a maximal ideal in a ring S, then S/J is a simple ring.

25 Let $\theta: S \to T$ be a ring homomorphism and let U be a subring of S. Prove that $\theta(U) = \{\theta(u) \mid u \in U\}$ is a subring of T.

26 Let J be any ideal of a ring S, and let U be a subring of S. Prove that $U \cap J$ is an ideal of U. Prove also that $U/U \cap J$ is isomorphic to a subring of T. [Let $v: S \to S/J$ be the natural epimorphism (p. 139). Define $\theta: U \to S/J$ to be the restriction of v to U, i.e. let $\theta(u) = v(u)$, all $u \in U$. Now use the Homomorphism Theorem of section 8.9.]

27 Let S be a commutative ring, $g(X) \in S[X]$ a polynomial of degree $m \geq 1$, and let $J = g(X)F[X]$ (J is the principal ideal of $S[X]$ generated by $g(X)$ – see section 8.11). For any $f(X) \in F[X]$, let $[f(X)]$ denote the coset $f(X) + J$. Prove that the map $\theta: S \to S/J$ given by the rule $\theta(c) = [c]$, all $c \in A$, is a ring monomorphism. Deduce that $\{[c] \mid c \in S\}$ is a subring of S/J which is isomorphic to S under the map $c \to [c]$.

28 Same notation as in the last exercise. Show that the ring S/J contains an element ξ which satisfies the equation $[g_m]\xi^m + [g_{m-1}]\xi^{m-1} + \cdots + [g_1]\xi + [g_0] = [0]$.

29 Show that $20Z \cap 12Z = 60Z$ (see Example 203, p. 140). Generalize this as follows: let a, b be non-zero elements of a principal ideal domain S. Show that $aZ \cap bZ = mZ$, where m is a *least common multiple* (lcm) of a and b, i.e. m satisfies **LCM1** $a \mid m$, $b \mid m$, and **LCM2**. If e is a non-zero element of S such that $a \mid e$, $b \mid e$, then $m \mid e$.

30 (S is any commutative ring.) Find quotient and remainder when $X^5 - X^3 + 1$ is divided by $X^2 + 1$.

31 Show that -2 is a zero of the polynomial $f(X) = X^3 + X^2 - 3X - 2$. Verify that $f(X) = (X + 2)h(X)$ for some $h(X)$ of degree 2.

32 *Interpolation problem.* Given three distinct elements a_1, a_2, a_3 of a field F, and given also elements b_1, b_2, b_3 of F (not necessarily distinct from each other, or from the a_i's), to find a polynomial $f(X) \in F[X]$ of degree 2, such that $f(a_1) = b_1$, $f(a_2) = b_2, f(a_3) = b_3$. [Take $g_1(X) = (X - a_2)(X - a_3)$, $g_2(X) = (X - a_1)(X - a_3), g_3(X) = (X - a_1)(X - a_2)$. Try to find elements $c_1, c_2, c_3 \in F$ so that $f(X) = \Sigma_{i=1}^3 c_i g_i(X)$ has the required properties. The result is called *Lagrange's Interpolation Formula.*]

33 Let p be a prime number and let $F = Z/pZ$, which is a field (p. 121). Write the elements $E_0, E_1, \ldots, E_{p-1}$ as $0, 1, \ldots, p - 1$, but remember that all calculations are to be done mod p.

Show that $a^{p-1} = 1$, for all $a \in F - \{0\}$. [Use the fact that $F - \{0\} = U(F)$ is a group of order $p - 1$.]

34 Same notation as in the last exercise. Show that *every* element a of F is a zero of the polynomial $X^p - X \in F[X]$. Deduce that $X^p - X = X(X - 1)(X - 2) \ldots (X - (p - 1))$. By comparing coefficients of X on the two sides of this equation, prove that $1.2.3 \ldots (p - 1) \equiv -1 \bmod p$ (Wilson's Theorem).

35 Prove that the only irreducible monic polynomials in $R[X]$ are (i) $X - a (a \in R)$, and (ii) $X^2 + aX + b$, where $a, b \in R$ satisfy $a^2 < 4b$.

36 Let $F = Z/2Z$. Find all irreducible polynomials in $F[X]$, of degrees 1, 2, 3 and 4.

37 Find a hcf d of $a = 420, b = 273$. Find also integers x, y such that $xa + yb = d$.

38 Find the inverse of $[45]$ in the field $Z/61Z$.

39 Find a hcf $d(X)$ of $a(X) = X^3 - 1, b(X) = X^4 + X^2 + 1$, elements of $R[X]$. Find also $g(X), h(X) \in R[X]$ such that $g(X)a(X) + h(X)b(X) = d(X)$.

40 Show that if $F = Z/2Z$, then $F[X]/(X^2 + X + 1)F[X]$ is a field of four elements. Write out the addition and multiplication tables of this field.

9 Vector spaces and matrices

9.1 Vector spaces over a field

In this chapter we shall study a new kind of algebraic structure, which is of great importance in many applications of mathematics. We start by choosing a field, which we denote by F; this will be kept fixed throughout the chapter. In many applications, F is either the field R of real numbers, or the field C of complex numbers. The only facts which we shall need from Chapter 8, are the definition and elementary properties of fields (see sections 8.1 and 8.4). The reader who has not read any of Chapter 8 can still follow the present chapter very well by taking F to be R throughout.

DEFINITION. Let F be a field. A *vector space over* F is a triple $(V, +, .)$, where V is a set, and

$$+ : V \times V \to V, \qquad . : F \times V \to V$$

are maps. The map $+ : V \times V \to V$ is a binary operation on V called *addition*; it takes each pair (v, w), with $v, w \in V$, to an element $v + w \in V$. The map $. : F \times V \to V$ is a kind of multiplication which takes each pair (α, v), with $\alpha \in F, v \in V$, to an element $\alpha v \in V$. Finally the following eight *vector space axioms* **FV1**–**FV8** must hold. It is convenient to collect these into two 'blocks' I and II.

I. These axioms involve only V and the addition operation $+$; they say that $(V, +)$ is an Abelian (i.e. commutative) group.

FV1 $(v + w) + x = v + (w + x)$,   for all $v, w, x \in V$.

FV2   There is an element $\mathbf{0}$ of V, such that $v + \mathbf{0} = v = \mathbf{0} + v$, for all $v \in V$.

FV3   To each $v \in V$ there is an element $-v \in V$, such that $v + (-v) = \mathbf{0} = (-v) + v$.

FV4 $v + w = w + v$,   for all $v, w \in V$.

II. These axioms involve F and the 'product' map $. : F \times V \to V$, as well as the additive group $(V, +)$.

FV5 $\alpha(\beta v) = (\alpha\beta)v$, for all α, $\beta \in F$ and $v \in V$.

FV6 $1_F v = v$, for all $v \in V$.

Here 1_F denotes the identity element of the field F.

FV7 $(\alpha + \beta)v = \alpha v + \beta v$, for all α, $\beta \in F$ and $v \in V$.

FV8 $\alpha(v + w) = \alpha v + \alpha w$, for all $\alpha \in F$ and v, $w \in V$.

Remarks

The definition just given is very similar to the definition of a ring (see section 8.1). Like a ring S, a vector space V has two 'operations', denoted as addition and multiplication. The first four axioms for a vector space correspond exactly to the first four axioms for a ring – they just say that $(V, +)$, like $(S, +)$, is an Abelian group. And the remaining axioms **FV5–FV8** look very much like axioms **R5–R8**. But the 'multiplication operation' $. : F \times V \to V$ for a vector space is of a different kind from the multiplication $. : S \times S \to S$ in a ring, because it brings in a new set F. In fact this operation $. : F \times V \to V$ is similar to the operation $* : G \times A \to A$ introduced in section 5.5; it 'combines' or 'multiplies' an element α of F with an element $v \in V$, to give a 'product' αv which is an element of V – we can say that the field F *acts on* V, in the same way that a group G may act on a set.

Scalars

It is useful to have a name to describe the elements of F, when we are talking about a vector space V over F. So an element α of F is called a *scalar*; scalars are usually not elements of V. The operation $. : F \times V \to V$ is sometimes called the *scalar action of F on V*, or the *multiplication by scalars*. It is usual to refer to a vector space $(V, +, .)$ simply as V.

We shall give a number of examples of vector spaces in the next section. For the moment, we should note some immediate consequences of the definition. Because $(V, +)$ is a group, the element **0** is unique; we use **0** (written or typed as $\underline{0}$) to distinguish this from the zero element $0 = 0_F$ of F. All the usual conventions for an additive Abelian group apply to V (see p. 52); for example the additive 'inverse' of $v \in V$ is written $-v$; we write $v - w$ to denote $v + (-w)$; if n is a positive integer and $v \in V$, then nv stands for $v + v + \ldots + v$ (n terms). There hold also the following *elementary theorems on vector spaces*.

THEOREM. *Let V be a vector space over the field F. Then*

(i) $(\alpha - \beta)v = \alpha v - \beta v$ *and* $\alpha(v - w) = \alpha v - \alpha w$,   *for all* $\alpha, \beta \in F$ *and* $v, w \in V$.

(ii) $\alpha \mathbf{0} = \mathbf{0} = 0_F v$,   *for all* $\alpha \in F$ *and* $v \in V$.

(iii) $(-\alpha)v = -\alpha v = \alpha(-v)$,   *for all* $\alpha \in F$ *and* $v \in V$.

(iv) *Let* $\alpha, \alpha_1, \ldots, \alpha_m$ *be elements of F, and* $v, v_1, \ldots, v_n$ *be elements of V. Then*

(a) $(\alpha_1 + \ldots + \alpha_m)v = \alpha_1 v + \ldots + \alpha_m v$,

(b) $\alpha(v_1 + \ldots + v_n) = \alpha v_1 + \ldots + \alpha v_n$ *and*

(c) $(\alpha_1 + \ldots + \alpha_n)(v_1 + \ldots + v_n) = \sum\limits_{i=1}^{m} \sum\limits_{j=1}^{n} \alpha_i v_j.$

The proofs are very similar to those of theorems (i)–(iv), section 8.2. So we leave most of these proofs as exercises (see the next example for the method).

Example 227. To prove theorem (i), notice that **FV7** gives $(\alpha - \beta)v + \beta v = [(\alpha - \beta) + \beta]v = \alpha v$. Add $-\beta v$ to both sides of this equation; we find $(\alpha - \beta)v = \alpha v - \beta v$. Next, **FV8** gives $\alpha(v - w) + \alpha w = \alpha[(v - w) + w] = \alpha v$. Add $-\alpha w$ to both sides. We get $\alpha(v - w) = \alpha v - \alpha w$. This completes the proof of (i). To prove $\alpha \mathbf{0} = \mathbf{0}$, just put $v = w$ in the last equation proved. To prove $0_F v = \mathbf{0}$, put $\alpha = \beta$ in $(a - b)v = \alpha v - \beta v$. Thus (ii) is proved.

Example 228. S-modules. The reader may notice that our definition of a vector space works just as well when the field F is replaced by an arbitrary ring S, because the definition does not use the fact that F is commutative, nor that every non-zero $\alpha \in F$ has an inverse. The algebraic structure one gets by replacing F by an arbitrary ring S in the definition of a vector space, i.e. a 'vector space over S', is usually called an *S-module*. Thus we could refer to a vector space over F as an *F-module*.

9.2 Examples of vector spaces

We give next the most important example of a vector space. As always, F is a field. Let n be a positive integer. Then a *vector of length n over F*, or more briefly an *n-vector over F*, is, by definition, an n-tuple

$$x = (x_1, \ldots, x_n)$$

whose *components* $x_1, \ldots, x_n$ are elements of F. It is not required that $x_1, \ldots, x_n$ be all distinct. The n-vectors $x = (x_1, \ldots, x_n)$, $y = (y_1, \ldots, y_n)$ are said to be *equal* (and we write $x = y$) if and only if

$$x_1 = y_2, \qquad x_2 = y_2, \ldots, x_n = y_n.$$

Given any n-vectors $x = (x_1, \ldots, x_n)$ and $y = (y_1, \ldots, y_n)$, we define their *sum* to be the n-vector

$$x + y = (x_1 + y_1, \ldots, x_n + y_n).$$

If α is any scalar (i.e. α is any element of F) we define

$$\alpha x = (\alpha x_1, \ldots, \alpha x_n).$$

THEOREM. *Let F be a field, and n any positive integer. Then the set $V = F^n$ of all n-vectors over F becomes a vector space over F, when we define the sum $x + y$ and scalar action αx as above. The zero element of F^n is the zero vector $(0, \ldots, 0)$. The negative of a vector $x = (x_1, \ldots, x_n)$ is $-x = (-x_1, \ldots, -x_n)$.*

Proof. This consists in checking the vector space axioms **FV1–FV8**, which is straightforward, and is left to the reader.

Vector spaces of type F^n, whose elements are vectors, are particularly important. But we shall see by the next two examples that there are vector spaces whose elements are *not* vectors – this is a little confusing at first! It would be better if a different term (such as 'F-module', see Example 228) were used to describe an algebraic structure $V = (V, +, .)$ as defined in section 9.1. But we shall follow the established practice, and refer to any such $(V, +, .)$ as a 'vector space', even though the elements of V are not necessarily vectors.

Example 229. Let $F[X]$ be the set of all polynomials in X with coefficients in F (see section 8.5). Then $F[X]$ is a vector space over F, if we take the sum of polynomials $f=f_0+f_1X+f_2X^2+\dots$ and $g=g_0+g_1X+g_2X^2+\dots$ to be $f+g=(f_0+g_0)+(f_1+g_1)X+\dots$ as usual, while if $\alpha\in F$, we define $\alpha f=\alpha f_0+(\alpha f_1)X+(\alpha f_2)X^2+\dots$. It is easy to check that $V=F[X]$, with these operations of addition and scalar action, satisfies **FV1**–**FV8**.

Example 230. Let A be any set, and let V be the set of all maps $f:A\to F$ (F being a field, as always). Then V becomes a vector space over F, if the operations of addition and scalar action are defined as follows. If $f,g\in V$, we define $f+g\in V$ by the rule: $(f+g)(a)=f(a)+g(a)$, for all $a\in A$. If $\alpha\in F$ and $f\in V$, define $\alpha f\in V$ by the rule: $(\alpha f)(a)=\alpha f(a)$, for all $a\in A$. Notice that $f+g$ and αf, as just defined, *are* elements of V, i.e. they are both maps of A into F. This is because $f(a)+g(a)$ and $\alpha f(a)$ are both elements of F – the sum and product being those which come from the fact that F is a ring. In checking axioms **FV1**–**FV8** for V, we have to use that the ring axioms **R1**–**R8** hold for F. As example, let us check **FV6**. We must prove that $1_F f=f$, for all $F\in V$. Putting $\alpha=1_F$ in our definition of αf, we have $(1_F f)(a)=1_F f(a)$. But $f(a)$ is an element of F, hence $1_F f(a)=f(a)$, by axiom **R6** applied to the ring F. Therefore $(1_F f)(a)=f(a)$, for all $a\in A$. Thus $1_F f=f$, which is what we wanted to prove. Verification of the other vector space axioms is carried out in the same way.

Example 231. Let n be a positive integer, and take $A=Z(n)=\{1,\dots,n\}$ in the last example. A map $f:Z(n)\to F$ is completely known when we know $f(1),f(2),\dots,f(n)$, which are elements of F. Therefore f is completely specified by the *vector* $(f(1),\dots,f(n))$, which is an element of F^n; we can regard $(f(1),\dots,f(n))$ simply as a notation for f. In this way, F^n can be regarded as the set of all maps $f:Z(n)\to F$. Then the definitions of $f+g$ and αf which were given in Example 230, coincide with the corresponding operations for n-vectors as given on p. 168.

9.3 Two geometric interpretations of vectors

Many features of vector space theory are best understood by giving vectors a geometric interpretation or illustration; conversely vectors often provide a very convenient way to solve a geometric problem.

In this section we describe two different (but closely related) interpretations of vectors. In both cases, the field F will be the real field R.

1. Vectors interpreted as points

In this interpretation, each n-vector $x = (x_1, \ldots, x_n) \in R^n$ represents the point in n-dimensional space having coordinates $(x_1, \ldots, x_n)$ relative to some given coordinate system – in particular, the zero vector $\mathbf{0} = (0, \ldots, 0)$ represents the origin of coordinates. Thus $x = (x_1, x_2)$ represents the point (x_1, x_2) in a plane (see Figure 18); an element $x = (x_1, x_2, x_3)$ of R^3 represents a point in ordinary space, and so on – of course it becomes difficult for ordinary people to visualize n-dimensional space if $n \geq 4$, and most mathematicians are content to consider R^n itself as the best formal definition of n-dimensional space. But for $n = 2$ and $n = 3$, where pictures can be drawn, one can 'illustrate' the vector space operations in a satisfactory way. From Figure 18 we see that the sum $x + y$ of two-vectors appears as the fourth vertex of the parallelogram having $\mathbf{0}x$ and $\mathbf{0}y$ as two of its sides: this is called the 'parallelogram law of vector addition'. If α is a scalar, i.e. if $\alpha \in R$, then $\alpha x = (\alpha x_1, \alpha x_2)$ represents a point on the line $\mathbf{0}x$ (we must assume that $x \neq (0, 0)$ for this to make sense), and any point on that line is given by αx for some $\alpha \in R$; for example $-x = (-1)x$ is the point on $\mathbf{0}x$, at the same distance from $\mathbf{0}$ as x, but on the opposite direction (see Figure 18).

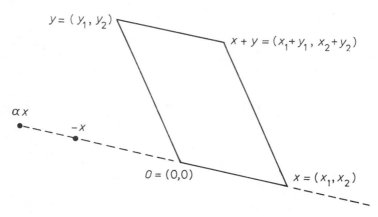

Figure 18

2. Vectors interpreted as translations

In this interpretation, each *n*-vector $x = (x_1, \ldots, x_n) \in R^n$ gives rise to a *map* $T_x : R^n \to R^n$, called *translation by x*, and defined by the rule

$$T_x(a) = x + a, \qquad \text{for all } a \in R^n. \tag{9.1}$$

Figure 19 illustrates such a map T_x, in case $n = 2$. Elements *a*, *b*, *c*, ... of R^2 are represented as points in the plane (as in Interpretation (9.1) above). The effect of T_x on a point *a* is to 'carry' or 'translate' the point along the line segment $a \to x + a$; these segments $a \to x + a$, $b \to x + b$, $c \to x + c$, ... are all parallel to and of the same length as the segment $\mathbf{0}x$ (which is the segment $\mathbf{0} \to x + \mathbf{0}$). (This was the original meaning of the word 'vector', meaning an agent which carries or 'conveys' points.)

If $x, y \in R^n$, it is easy to calculate the product of the maps T_x, T_y. We have from (9.1) that $T_x[T_y(a)] = x + T_y(a) = x + y + a$, for all $a \in R^n$. But this means that $T_x T_y$ has the same effect, on any $a \in R^n$, as the translation T_{x+y}; in other words $T_x T_y = T_{x+y}$. Since the sum of vectors is commutative, we have $T_{x+y} = T_{y+x}$, hence

$$T_x T_y = T_{x+y} = T_y T_x, \qquad \text{for all } x, y \in R^n. \tag{9.2}$$

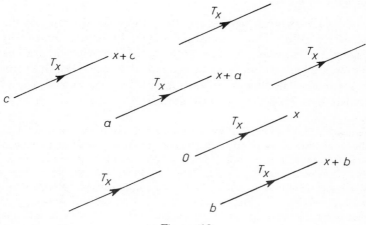

Figure 19

Example 232. Figure 20 illustrates equations (9.2) geometrically in the case $n = 2$. If we start from a point a in R^2 and apply first T_y and then T_x, we are following the path $a \rightarrow y + a \rightarrow x + y + a$, i.e. we follow the two left-hand sides of the parallelogram. The result is the same as if we had applied T_{x+y} to a, and since this is true for all a, we have $T_y T_x = T_{y+x}$. If we apply first T_x and then T_y to a, we follow the two right-hand sides of the parallelogram. We get to the same point $x + y + a$ as before; this shows that $T_y T_x = T_x T_y$.

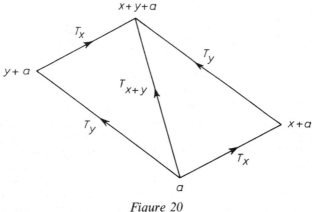

Figure 20

Remark on notation

Many authors prefer to represent the translation T_x simply by the vector x. With this convention Figure 20 would appear as Figure 21 (the *points* of the figure, since they cannot now be represented as vectors, have been denoted A, B, C, D). But from now on we shall always represent points by vectors, and denote translations T_x, T_y, etc., as in Figure 20.

Example 233. Lines in R^n. If we interpret R^n as n-dimensional space, and the elements of R^n as points, then other geometrical objects can be defined as *subsets of R^n*. For example if x is any non-0 vector in R^n, we define *the line through 0 and x* to be the set $L(0, x) = \{\lambda x | \lambda \in R\}$. This purely 'algebraic' definition is in accord with our common sense, at least when $n = 2$. For we see from Figure 22 that if we take a fixed two-vector x, and then mark the points λx, as λ takes various real values, then these points will range over the straight line through 0 and x. Also we find that our

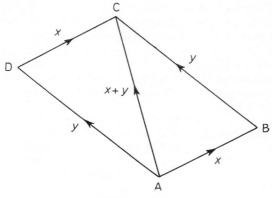

Figure 21

'lines' $L(\mathbf{0}, x)$ have another property which we should expect, if these are to resemble the straight lines of ordinary experience, namely if y is any point of $L(\mathbf{0}, x)$, except $\mathbf{0}$ itself, then the line $L(\mathbf{0}, y)$ through $\mathbf{0}$ and y, is the same as the line $L(\mathbf{0}, x)$ (see Figure 22). To prove this, we have to prove that the *sets* $L(\mathbf{0}, x)$ and $L(\mathbf{0}, y)$ are equal. This is quite easy: we know that $y = \beta x$ for some $\beta \in R$, because $y \in L(\mathbf{0}, x)$. Hence for any $\lambda \in R$, the vector $\lambda y = \lambda(\beta x) = (\lambda \beta) x \in L(\mathbf{0}, x)$; this proves that $L(\mathbf{0}, y) \subseteq L(\mathbf{0}, x)$. But the

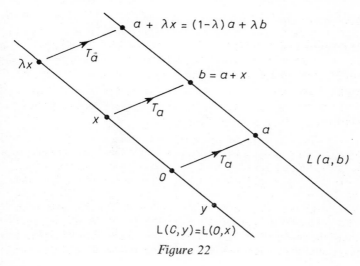

Figure 22

scalar β is not zero, because $y = \beta x$ is not $\mathbf{0}$. So $x = \beta^{-1}(\beta x) = \beta^{-1} y$, and we find by the same argument as above that $L(\mathbf{0}, x) \subseteq L(\mathbf{0}, y)$. Therefore $L(\mathbf{0}, x) = L(\mathbf{0}, y)$, which is what we wanted to prove.

Example 234. Line through two given points. So far we have dealt only with lines which contain $\mathbf{0}$. We should like to define the line $L(a, b)$ through any given distinct points $a, b \in R^n$. We may do this as follows: let $x = b - a$ (this is not $\mathbf{0}$, since $a \neq b$ by assumption), and then apply the translation map T_a to the line $L(\mathbf{0}, x)$. T_a takes $\mathbf{0}$ to a, it takes x to $a + x = b$, and it takes the general point λx of $L(\mathbf{0}, x)$ to $a + \lambda x = a + \lambda(b - a) = (1 - \lambda)a + \lambda b$ (see Figure 22). So it is reasonable to define the line $L(a, b)$ to be the set $\{(1 - \lambda)a + \lambda b \,|\, \lambda \in R\}$. It is interesting to take different values of λ, and see where the points $(1 - \lambda)a + \lambda b$ lie; for example $\lambda = 0$ gives a, $\lambda = 1$ gives b, $\lambda = \frac{1}{2}$ gives the mid-point $\frac{1}{2}(a + b)$ of the segment ab; points inside this segment correspond to values of λ satisfying $0 < \lambda < 1$.

9.4   Subspaces

We return now to the general theory of vector spaces, and define *subspaces* of a given vector space; these are like the subgroups of a group (section 5.2), or the subrings of a ring (section 8.2).

DEFINITION. Let V be a vector space over a field F. Then a non-empty subset U of V is called a *subspace of V* if it satisfies the following two conditions:

    SS1 $x, y \in U \Rightarrow x + y \in U$,

and

    SS2 $x \in U, \lambda \in F \Rightarrow \lambda x \in U$.

It follows from this definition that a subspace U of V is always a *subgroup* of the additive group $(V, +)$. For U satisfies the subgroup condition **S1** (p. 58), since this is the same as **SS1**. But U also satisfies **S2**, as we see by taking $\lambda = -1$ in **SS2**; hence U is a subgroup of $(V, +)$. From this fact (or by taking $\lambda = 0$ in **SS2** and using Theorem (ii), section 9.1), we see that *a subspace U always contains the element* $\mathbf{0}$. The one-element set $\{\mathbf{0}\}$ is a subspace of V, and V itself is a subspace of V. Notice that a subspace U of V is itself a vector space over F; the vector space axioms **FV1–FV8** hold for U, because they hold for V.

Example 235. Let Z^2 denote the set of elements (z_1, z_2) of R^2 whose components z_1, z_2 are both integers. This is a *subgroup* of the additive group $(R^2, +)$, as we see by checking the subgroup conditions **S1, S2** (p. 58). But it is not a *subspace* of R^2 since it does not satisfy **SS2**: for example if $x = (1, 1)$ and $\lambda = \frac{1}{3}$, it is clear that $x \in Z^2$ but $\lambda x \notin Z^2$.

Example 236. Let p be a given element of a vector space V, and define $U = \{\alpha p | \alpha \in F\}$. Then U is a subspace of V. For if $x, y \in U$ we must have $x = \alpha p$, $y = \beta p$, for some elements α, $\beta \in F$. Then $x + y = \alpha p + \beta p = (\alpha + \beta)p \in U$, which shows that U satisfies **SS1**. If λ is any element of F, then $\lambda x = \lambda(\alpha p) = (\lambda \alpha)p \in U$, which shows that U satisfies **SS2**.

Notice that if $F = R$ and $V = R^n$, and if $p \neq \mathbf{0}$, then U is the line $L(\mathbf{0}, p)$ defined in Example 233: this shows that *every line through* $\mathbf{0}$ *is a subspace of* R^n. But a line $L(a, b)$ which does not contain $\mathbf{0}$ cannot be a subspace, since a subspace always contains $\mathbf{0}$.

Example 237. Let $a = (a_1, \ldots, a_n)$ be a given element of the vector space F^n. Define U to be the set of all $x \in R^n$ which satisfy the equation $a_1 x_1 + \ldots + a_n x_n = 0$. Then U is a subspace of F^n. For if $x, y \in F^n$ both satisfy this equation, then $a_1(x_1 + y_1) + \ldots + a_n(x_n + y_n) = (a_1 x_1 + \ldots + a_n x_n) + (a_1 y_1 + \ldots + a_n y_n) = 0$, i.e. $x + y$ satisfies the equation too. This shows that $x, y \in U \Rightarrow x + y \in U$. And if λ is any element of F, then $x \in U \Rightarrow \lambda x \in U$, since $a_1(\lambda x_1) + \ldots + a_n(\lambda x_n) = \lambda(a_1 x_1 + \ldots + a_n x_n) = 0$.

Example 238. Let A be a subset of the real number field R, and let V be the set of all maps ($=$ functions) $f : A \to R$. We saw in Example 230 how to make V into a vector space over R. Now let U be the set of all *continuous* functions $f : A \to R$. From elementary analysis we know that if f, g are continuous functions then so is $f + g$, and so also is λf, for any real number λ. Therefore U is a subspace of V.

We end this section with a general theorem about subspaces.

THEOREM. *Let* U, U' *be subspaces of a vector space V over F. Then* (i) $U \cap U'$ *is a subspace of V, and* (ii) $U + U'$ *is a subspace of V, where* $U + U'$, *which is called the 'sum' of U and U', is defined by*

$$U + U' = \{u + u' | u \in U, u' \in U'\}. \tag{9.3}$$

Proof. (i) Let x, y be elements of $U \cap U'$, and λ any element of F. Then $x + y$ and λx belong to U, because x, $y \in U$ and U is a subspace. Similarly $x + y$, $\lambda x \in U'$. But this shows that $x + y$, $\lambda x \in U \cap U'$ satisfies **SS1** and **SS2**, as required.

(ii) Let x, y be elements of $U + U'$, and λ any element of F. By the definition of $U + U'$ we have $x = u + u'$, $y = v + v'$, for some u, $v \in U$ and u', $v' \in U'$. Therefore $x + y = (u + u') + (v + v') = (u + v) + (u' + v') \in U + U'$, since $u + v \in U$, $u' + v' \in U'$. This shows that $U + U'$ satisfies **SS1**. To show $U + U'$ satisfies **SS2**, we work out $\lambda x = \lambda(u + u') = \lambda u + \lambda u'$; this proves that $\lambda x \in U + U'$, since $\lambda u \in U$ and $\lambda u' \in U'$. This completes the proof of the theorem.

The definition (9.3) of the sum of two subspaces U, U' of V, can be extended to define the sum $U_1 + \ldots + U_n$ of any $n(\geq 2)$ subspaces $U_1, \ldots, U_n$ of V, namely we put

$$U_1 + \ldots + U_n = \{u_1 + \ldots u_n | u_1 \in U_1, \ldots, u_n \in U_n\}. \tag{9.4}$$

It can then be proved that $U_1 + \ldots + U_n$ is a subspace of V, in the same way as we proved that $U + U'$ is a subspace in the theorem above. Notice that this sum of subspaces is commutative and associative; we have $U_1 + U_2 = U_2 + U_1$ and $(U_1 + U_2) + U_3 = U_1 + (U_2 + U_3) = U_1 + U_2 + U_3$, for any subspaces U_1, U_2, U_3 of V.

9.5 Linear combinations. Spanning sets

Suppose that V is a vector space over F, and that $x_1, \ldots, x_n$ are some elements of V. Then any element of V which can be written in the form

$$\lambda_1 x_1 + \ldots + \lambda_n x_n, \tag{9.5}$$

where $\lambda_1, \ldots, \lambda_n \in F$, is called a *linear combination* of $x_1, \ldots, x_n$. The scalars $\lambda_1, \ldots, \lambda_n$ are called the *coefficients* of (9.5). The set of all linear combinations (9.5) of the given elements $x_1, \ldots, x_n$ of V, is denoted

$$\text{sp}\{x_1, \ldots, x_n\} \qquad \text{or sp } S, \tag{9.6}$$

the second notation being used if it is convenient to regard $x_1, \ldots, x_n$ as forming a set $S = \{x_1, \ldots, x_n\}$. The fundamental property of $\mathrm{sp}\{x_1, \ldots, x_n\}$ is that *it is a subspace of V*, and in fact it is the smallest subspace of V which contains the given elements $x_1, \ldots, x_n$, as the next theorem shows.

THEOREM. *Let $x_1, \ldots, x_n$ be elements of a vector space V, then* (i) $\mathrm{sp}\{x_1, \ldots, x_n\}$ *is a subspace of V, and* (ii) *If U is any subspace of V which contains $x_1, \ldots, x_n$, then $U \supseteq \mathrm{sp}\{x_1, \ldots, x_n\}$.*

Proof. (i) Take any elements u, v of the set $\mathrm{sp}\{x_1, \ldots, x_n\}$, so that $u = \alpha_1 x_1 + \ldots + \alpha_n x_n$ and $v = \beta_1 x_1 + \ldots \beta_n x_n$, for some $\alpha_1, \ldots, \alpha_n, \beta_1, \ldots, \beta_n \in F$. Then $u + v = (\alpha_1 + \beta_1)x_1 + \ldots + (\alpha_n + \beta_n)x_n$, which shows that $u + v \in \mathrm{sp}\{x_1, \ldots, x_n\}$. And if λ is any element of F we have $\lambda u = (\lambda\alpha_1)x_1 + \ldots + (\lambda\alpha_n)x_n$, which shows that $\lambda u \in \mathrm{sp}\{x_1, \ldots, x_n\}$. Thus $\mathrm{sp}\{x_1, \ldots, x_n\}$ satisfies both subspace conditions **SS1** and **SS2**.

(ii) Since U satisfies **SS2** and $x_1 \in U$, it follows that $\lambda_1 x_1 \in U$, for any $\lambda_1 \in F$; similarly, U contains $\lambda_2 x_2, \ldots, \lambda_n x_n$, for any $\lambda_2, \ldots, \lambda_n \in F$. But since U satisfies **SS1** it also contains the sum of $\lambda_1 x_1, \ldots, \lambda_n x_n$, i.e. it contains every element (9.5) of the set $\mathrm{sp}\{x_1, \ldots, x_n\}$.

DEFINITION. $\mathrm{sp}\{x_1, \ldots, x_n\} = \mathrm{sp} \ S$ is called the subspace of V *spanned* (or *generated*) by $x_1, \ldots, x_n$, or by the set $S = \{x_1, \ldots, x_n\}$. If $V = \mathrm{sp}\{x_1, \ldots, x_n\} = \mathrm{sp} \ S$, we say that $S = \{x_1, \ldots, x_n\}$ is a *spanning set* (or a *set of generators*) for V.

Remark

$\mathrm{sp} \ S$ is the analogue, for vector spaces, of the set $\mathrm{gp} \ X$ introduced in section 5.3, where X was any non-empty subset of a group G. The elements of $\mathrm{gp} \ X$ are all the 'words' in X, i.e. all the elements of G you can get by applying the group operations (multiplication and taking inverses) to the elements of X. Similarly $\mathrm{sp} \ S$ is the set of all the elements of V which you can get by applying the vector space operations (addition and multiplication by scalars) to the elements of S, i.e. it is the set of all linear combinations of these elements.

Example 239. If S has only one element p, then $\mathrm{sp} \ S = \mathrm{sp}\{p\}$ is the set

$\{\alpha p | \alpha \in F\}$. We showed in Example 235 that this is a subspace of V. If $p = \mathbf{0}$, the $\mathrm{sp}\{p\} = \{\mathbf{0}\}$.

Example 240. Suppose S has two elements p, q. Then sp S consists of all elements $\alpha p + \beta q$, for α, $\beta \in F$. If q is a scalar multiple of p, i.e. if $q = \delta p$ for some $\delta \in F$ (and this is equivalent to saying that $q \in \mathrm{sp}\{p\}$), then $\mathrm{sp}\{p, q\} = \mathrm{sp}\{p\}$. For certainly $\mathrm{sp}\{p\} \subseteq \mathrm{sp}\{p, q\}$ (this is true for any p, $q \in V$), but also $\mathrm{sp}\{p, q\} \subseteq \mathrm{sp}\{p\}$, since any linear combination $\alpha p + \beta q = (\alpha + \beta \delta)p$ lies in $\mathrm{sp}\{p\}$. Similarly, if $p \in \mathrm{sp}\{q\}$, which means that p is a scalar multiple of q, then $\mathrm{sp}\{p, q\} = \mathrm{sp}\{q\}$. These are special cases of a general principle which we shall use in the next section: if one of the elements, say x_1, of the set $S = \{x_1, \ldots, x_n\}$ belongs to the subspace $\mathrm{sp}\{x_2, \ldots, x_n\}$, then $\mathrm{sp}\{x_1, x_2, \ldots, x_n\} = \mathrm{sp}\{x_2, \ldots, x_n\}$, i.e. x_1 can be removed from S with altering the space it spans.

Example 241. Planes in R^n. In Example 233 we defined the line through $\mathbf{0}$ and x, where x is any non-$\mathbf{0}$ element of R^n, to be the set $L(\mathbf{0}, x) = \{\lambda x | \lambda \in R\}$; with our new notation we can write this definition: $L(\mathbf{0}, x) = \mathrm{sp}\{x\}$. Now we shall define the *plane through $\mathbf{0}$, x and y* to be the set $L(\mathbf{0}, x, y) = \mathrm{sp}\{x, y\}$. But in making this definition we cannot take *any* points x, $y \in R^n$, we must assume (i) that $y \notin \mathrm{sp}\{x\}$, and (ii) that $x \notin \mathrm{sp}\{y\}$. For the last example shows that if $y \in \mathrm{sp}\{x\}$ or if $x \in \mathrm{sp}\{y\}$, then $\mathrm{sp}\{x, y\}$ is equal to $\mathrm{sp}\{x\}$ or to $\mathrm{sp}\{y\}$, which shows that our so-called plane has degenerated into a line (or, in the extreme cases that $x = y = \mathbf{0}$, into the set $\{\mathbf{0}\}$). The combined conditions (i) and (ii) are equivalent to the condition that the points $\mathbf{0}$, x, y do not lie on a single line. Figure 23 illustrates the plane (in three-dimensional space R^3) through the points $\mathbf{0}$, $x = (1, 1, 1)$ and $y = (1, 1, 0)$. It is clear that neither $x = (1, 1, 1)$ nor $y = (1, 1, 0)$ is a scalar multiple of the other, so this plane is well defined.

If we want to define the plane through points a, b, c of R^n, none which is the point $\mathbf{0}$, we use the 'translation' trick of Example 234. Define $x = b - a$, $y = c - a$, and apply the translation T_a to the plane $L(\mathbf{0}, x, y)$. T_a takes $\mathbf{0}$ to a, x to b, y to c; it takes the general point $\lambda x + \mu y$ of $L(\mathbf{0}, x, y)$ to $a + \lambda(b - a) + \mu(c - a) = (1 - \lambda - \mu)a + \lambda b + \mu c$. So we define the plane $L(a, b, c)$ through a, b, c to be the set $\{(1 - \lambda - \mu)a + \lambda b + \mu c | \lambda, \mu \in R\}$. Of course we can do this only if x, y satisfy the conditions (i) and (ii) above, i.e. if $\mathbf{0}$, x, y do not lie on a single line, i.e. if a, b, c do not lie on a single line.

Example 242. The vectors $e_1 = (1, 0, 0, \ldots, 0)$, $e_2 = (0, 1, 0, \ldots, 0)$, $\ldots$, $e_n = (0, 0, 0, \ldots, 1)$ in F^n are called *unit vectors*. The set $\{e_1, e_2, \ldots, e_n\}$ is a spanning set for F^n, since any element $a = (a_1, a_2, \ldots, a_n)$ can be written

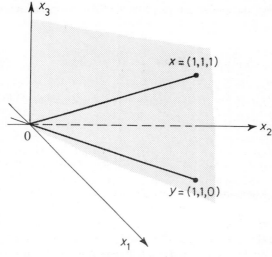

Figure 23

as $a_1e_1 + a_2e_1 + \ldots + a_ne_n$ (notice that here $a_1, a_2, \ldots, a_n$ are scalars), i.e. a is an element of $\mathrm{sp}\{e_1, \ldots, e_n\}$. Therefore $F^n \subseteq \mathrm{sp}\{e_1, \ldots, e_n\}$; but of course $\mathrm{sp}\{e_1, \ldots, e_n\} \subseteq F^n$, and so $F^n = \mathrm{sp}\{e_1, \ldots, e_n\}$. In general there are many spanning sets for a vector space. For example F^n has a spanning set $\{f_1, \ldots, f_n\}$, where $f_1 = e_1$, $f_2 = e_1 + e_2 = (1, 1, 0, \ldots, 0)$, $f_3 = e_1 + e_2 + e_3 = (1, 1, 1, \ldots, 0), \ldots f_n = e_1 + e_2 + \ldots + e_n = (1, 1, 1, \ldots, 1)$. The reader can check this by showing that any $a = (a_1, a_2, \ldots, a_n)$ can be written as a linear combination of $f_1, f_2, \ldots, f_n$.

Example 243. We saw in Example 229 that the ring $F[X]$ of all polynomials can be regarded as a vector space over F. But $F[X]$ has no *finite* spanning set S. For if $S = \{f_1(X), \ldots, f_n(X)\}$ is any finite subset of $F[X]$, let N be the maximum of the degrees of these polynomials $f_1(X)$, $\ldots, f_n(X)$. Clearly any linear combination $\lambda_1 f_1(X) + \ldots + \lambda_n f_n(X)$ has degree at most N. Therefore, for example, the polynomial X^{N+1} does not lie in sp S.

Example 244. We have defined sp S only for *finite* sets S, but there is no difficulty in extending the definition to cover an arbitrary subset S of a vector space V; we take sp S to be the set of all linear combinations $\lambda_1 s_1 + \ldots + \lambda_n s_n$ where $\{s_1, \ldots, s_n\}$ runs over all finite subsets of S. With this definition, the infinite set $S = \{1, X, X^2, X^3, \ldots\}$ is a spanning set for the vector space $V = F[X]$ of the last example.

9.6 Linear dependence. Basis of a vector space

Suppose that V is a vector space over F, and that $x_1, \ldots, x_n$ are elements of V. It may well happen that some linear combination of $x_1, \ldots, x_n$ turns out to be equal to the zero element $\mathbf{0}$ of V, say

$$\lambda_1 x_1 + \ldots + \lambda_n x_n = \mathbf{0}. \tag{9.7}$$

An equation of the form (9.7) is called a *linear dependence relation* on $x_1, \ldots, x_n$, or on the set $S = \{x_1, \ldots, x_n\}$. Of course there is always at least one such relation, *viz.*

$$0x_1 + \ldots + 0x_n = \mathbf{0}, \tag{9.8}$$

in which the coefficients $\lambda_1, \ldots, \lambda_n$ are all equal to the zero element 0 of F; this is called the *trivial relation*. A relation† (9.7) is *non-trivial*, therefore, if at least one of the coefficients $\lambda_1, \ldots, \lambda_n$ is not zero.

DEFINITION. The set $S = \{x_1, \ldots, x_n\}$ is called *linearly dependent* if there exists a non-trivial linear relation (9.7). It is called *linearly independent* if there is no non-trivial relation, i.e. if $\lambda_1 x_1 + \ldots + \lambda_n x_n$ is equal to $\mathbf{0}$ only if $\lambda_1 = \ldots = \lambda_n = 0$.

Example 245. Let $S = \{x, y, z\}$, where $x = (1, 0)$, $y = (1, 1)$, $z = (2, -1)$ are elements of $V = R^2$. We can find a non-trivial relation $3x - y - z = (3, 0) + (-1, -1) + (-2, 1) = (0, 0) = \mathbf{0}$, hence S is a linearly dependent set. But the set $\{x, y\}$ is independent. For if we have $\lambda x + \mu y = \mathbf{0}$, with λ, $\mu \in R$, then $(\lambda, 0) + (\mu, \mu) = (0, 0)$, i.e. $(\lambda + \mu, \mu) = (0, 0)$, which gives two equations $\lambda + \mu = 0$, $\mu = 0$. It is clear that these imply $\lambda = \mu = 0$. Hence the only relation $\lambda x + \mu y = \mathbf{0}$ is the trivial one. The reader can prove in a similar way that both of the sets $\{x, z\}$ and $\{y, z\}$ are linearly independent.

Example 246. Let V be any vector space. If one of the elements $x_1, \ldots, x_n$ is $\mathbf{0}$, then the set $S = \{x_1, \ldots, x_n\}$ is linearly dependent. For example if

---

†We usually shorten the expression 'linear dependence relation' to 'linear relation', or simply to 'relation'.

$x_1 = 0$, then we have the non-trivial relation $1x_1 + 0x_2 + \ldots + 0x_n = 0$. Putting this in another way; a *linearly independent set* $\{x_1, \ldots, x_n\}$ *never has* 0 *as a member*.

Example 247. Let x be an element of a vector space V. Then *the one-element set* $\{x\}$ *is linearly independent if and only if* $x \neq 0$. For (as we have just seen) if $x = 0$ then there is a non-trivial relation $1x = 0$, which shows $\{x\}$ is dependent. If $x \neq 0$ and if $\lambda x = 0$ is a linear relation on $\{x\}$, we must have $\lambda = 0$. For if $\lambda \neq 0$ there would exist an inverse $\lambda^{-1} \in F$, and $\lambda x = 0 \Rightarrow \lambda^{-1}(\lambda x) \Rightarrow 0 \Rightarrow (\lambda^{-1}\lambda)x = 0 \Rightarrow 1x = 0 \Rightarrow x = 0$, contradicting our assumption that $x \neq 0$. So the only linear relation $\lambda x = 0$ is the trivial one, hence $\{x\}$ is a linearly independent set.

Now consider a two-element subset $\{x, y\}$ of V. We shall show that $\{x, y\}$ is linearly independent if and only if it satisfies

$$x \notin \mathrm{sp}\{y\} \qquad \text{and} \qquad y \notin \mathrm{sp}\{x\}. \tag{9.9}$$

For if $x \in \mathrm{sp}\{y\}$, i.e. if $x = \alpha y$ for some $\alpha \in F$, there holds the non-trivial relation $1x + (-\alpha)y = 0$, showing that $\{x, y\}$ is a dependent set; we get the same conclusion if $y \in \mathrm{sp}\{x\}$. Thus if $\{x, y\}$ is independent, then (9.9) holds. Now assume (9.9) holds, and that $\lambda x + \mu y = 0$. If $\lambda \neq 0$, we find $x = -\lambda^{-1}\mu y$, i.e. x is a scalar multiple of y. But by (9.9) this cannot happen. Hence $\lambda = 0$. If $\mu \neq 0$ we have $y = -\mu^{-1}\lambda x$, and again this cannot happen. So $\mu = 0$. Thus the only relation $\lambda x + \mu y = 0$ is the trivial one. Therefore the set $\{x, y\}$ is linearly independent.

In case $V = R^n$ we have a *geometric interpretation of linear independence*. For by Example 241, together with what we have just proved, we can say that $\{x, y\}$ is *dependent* if and only if x lies on the line $L(0, y)$, or y lies on $L(0, x)$. So the plane $L(0, x, y)$ is well defined precisely when $\{x, y\}$ is an independent set, and this happens precisely when 0, x, y do not lie on a single line.

The next theorem shows that if $S = \{x_1, \ldots, x_n\}$ is a linearly dependent set, then one can remove one of its elements without altering the subspace $\mathrm{sp}\{x_1, \ldots, x_n\}$.

THEOREM. (i) *Let* $S = \{x_1, \ldots, x_n\}$ *be a linearly dependent subset of a vector space* V, *then there is (at least) one element* $x_i \in S$ *which is a linear combination of the other elements of* S. *In this case we have*

$$\mathrm{sp}\{x_1, \ldots, x_n\} = \mathrm{sp}\{x_1, \ldots, x_{i-1}, x_{i+1}, \ldots, x_n\}.$$

Proof. Since S is linearly dependent, there must be a non-trivial relation (9.7). Suppose that the coefficient $\lambda_i \neq 0$. Then from (9.7) we get $\lambda_i x_i = -\lambda_1 x_1 - \ldots - \lambda_{i-1} x_{i-1} - \lambda_{i+1} x_{i+1} \ldots - \lambda_n x_n$, hence multiplying both sides of this equation by λ_i^{-1}, and then writing $\alpha_j = -\lambda_i^{-1} \lambda_j (j = 1, \ldots, i-1, i+1, \ldots, n)$, we have

$$x_i = \alpha_1 x_1 + \ldots + \alpha_{i-1} x_{i+1} + \alpha_{i+1} x_{i+1} + \ldots + \alpha_n x_n, \qquad (9.10)$$

showing that x_i is a linear combination of the other elements of S (notice that we can do this for any x_i, which appears with non-zero coefficient λ_i in the relation (9.7)).

Suppose now that (9.10) holds. Write $U = \mathrm{sp}\{x_1, \ldots, x_{i-1}, x_{i+1}, \ldots, x_n\}$. Clearly $U \subseteq \mathrm{sp}\ S$, since any linear combination of $x_1, \ldots, x_{i-1}, x_{i+1}, \ldots, x_n$ can be regarded as a linear combination of $x_1, \ldots, x_n$. Conversely any element $v = \mu_1 x_1 + \ldots + \mu_i x_i + \ldots + \mu_n x_n$ of sp S can be re-written as a linear combination of $x_1, \ldots, x_{i-1}, x_{i+1}, \ldots, x_n$, if we use (9.10) to replace the x_i which appears in our expression for v. Hence $v \in U$. This proves that sp $S \subseteq U$, hence sp $S = U$, which is what we wanted to prove.

Example 248. In Example 245 we had a non-trivial relation $3x - y - z = \mathbf{0}$. We can use this to express any one of x, y, z as a linear combination of the other two; for example $x = \frac{1}{3}y + \frac{1}{3}z$. The theorem above then shows that $\mathrm{sp}\{x, y, z\} = \mathrm{sp}\{y, z\}$. Similarly we find $\mathrm{sp}\{x, y, z\} = \mathrm{sp}\{x, z\} = \mathrm{sp}\{x, y\}$. We shall see in Example 249 that all these spaces are equal to $V = R^2$.

Basis of a vector space

We come now to the fundamental definition of vector space theory.

DEFINITION. A finite subset $S = \{x_1, \ldots, x_n\}$ of a vector space V is called a *basis of V* if it satisfies the two conditions

B1   sp $S = V$, i.e. S is a spanning set for V,

and

B2 S is linearly independent.

Example 249. Consider the vectors x, y, z of Examples 245 and 247. We shall show that $\{x, y\} = \{(1, 0), (1, 1)\}$ is a basis for the vector space $V = R^2$. It is linearly independent, as we saw in Example 245. To show it is a spanning set for R^2, let $a = (a_1, a_2)$ be any element of R^2. We can find λ, $\mu \in R$ such that $\lambda x + \mu y = a$, for this means that $(\lambda, 0) + (\mu, \mu) = (a_1, a_2)$, i.e. $\lambda + \mu = a_1$ and $\mu = a_2$. These equations for λ, μ are easily solved, giving $\lambda = a_1 - a_2$, $\mu = a_2$. Hence $\mathrm{sp}\{x, y\}$ contains every element of R^2, and therefore $\mathrm{sp}\{x, y\} = R^2$. We have now proved that $\{x, y\}$ satisfies both conditions **B1** and **B2**, and so $\{x, y\}$ is a basis for R^2. In fact the sets $\{x, z\}$ and $\{y, z\}$ are also bases for R^2. For we saw from Example 248 that $\mathrm{sp}\{x, y, z\} = \mathrm{sp}\{x, z\} = \mathrm{sp}\{y, z\}$, and since $\mathrm{sp}\{x, y\} = R^2$, we have $\mathrm{sp}\{x, z\} = R^2$ and $\mathrm{sp}\{y, z\} = R^2$. We can check (see Example 245) that $\{x, z\}$ and $\{y, z\}$ are both linearly independent. So both these sets are bases of R^2. Notice that $\{x, y, z\}$ is *not* a basis of R^2, since although it spans R^2, i.e. it satisfies **B1**, it is not independent and so does not satisfy **B2**.

Example 250. Let $V = F^n$, the vector space of all n-vectors over F. We saw in Example 242 that unit vectors $e_1, \ldots, e_n$ form a spanning set for F^n. We shall now prove that the set $\{e_1, \ldots, e_n\}$ is linearly independent, hence it is a basis for F^n. Let $\lambda_1 e_1 + \ldots + \lambda_n e_n = \mathbf{0}$ be a linear relation. Since the left side of this equation is equal to $(\lambda_1, \ldots, \lambda_n)$, we have $(\lambda_1, \ldots, \lambda_n) = (0, \ldots, 0)$. But then $\lambda_1, \ldots, \lambda_n$ are all equal to zero, hence our relation is trivial; this proves that $\{e_1, \ldots, e_n\}$ is linearly independent.

A vector space may have (in fact usually does have) many bases: for example R^2 has the three bases given in Example 250, and also the basis of unit vectors $\{(1, 0), (0, 1)\}$, and infinitely many others.

Existence of bases

We must consider the question, whether a given vector space V has a basis? One necessary condition, immediate from the definition of basis, is that V should have a finite spanning set; moreover this is a genuine restriction, since we saw in Example 243 an example of a vector space which has no finite spanning set. We make the following definition.

DEFINITION. Let V be a vector space over F. We say that V is *finitely generated* (or *finitely spanned*) if there is a finite subset $S = \{x_1, \ldots, x_n\}$ of V such that sp $S = V$.

Suppose now that sp $S = V$, as just described. This set S already satisfies the condition **B1**; if S is also linearly independent, it satisfies **B2** and is a basis of V. But if S is linearly *dependent*, then by theorem (i) there is some element of S – and by renumbering $x_1, \ldots, x_n$ if necessary we can assume that this element is x_1 – which is a linear combination of the others; theorem (i) also shows that sp $S = \text{sp}\{x_2, \ldots, x_n\}$, so that $\{x_2, \ldots, x_n\}$ is still a spanning set for V. If this set is linearly independent, it satisfies **B1** and **B2**, and so is a basis of V. If it is dependent, we may remove another element (which we may assume is x_2), so that $\{x_3, \ldots, x_n\}$ still spans V. Continuing in this way we must eventually arrive at a subset S' of S which spans V *and* is linearly independent, i.e. which is a basis for V. This proves the next theorem.

THEOREM. (ii) *Let V be a vector space over F, and assume that V is finitely generated. Then V has a basis. In fact if S is any finite spanning set for V, there is a subset S' of S (which might be equal to S) which is a basis of V.*

Remark

It is possible that the above process will give the basis $S' = \varnothing$ – this happens if $V = \{\mathbf{0}\}$. We saw in Example 246 that no set containing $\mathbf{0}$ can be independent, so if $V = \{\mathbf{0}\}$, the empty subset of V is the only possible basis. We make it a convention, that $\{\mathbf{0}\}$ has basis $\varnothing$; this sounds strange, but in fact fits in with our general theory. We extend our definitions of spanning set and linear independence to the empty set, as follows: If $\varnothing$ is the empty subset of a vector space, then the set spanned by $\varnothing$ is defined to be $\{\mathbf{0}\}$; moreover $\varnothing$ is considered to be a linearly independent set.

9.7 The Basis Theorem. Dimension

Throughout this section V is assumed to be a *finitely generated* vector space over the field F. We know from theorem (ii) in the last section that V has at least one basis, and we also know from Example 250 that V may have many bases. The main statement of the Basis Theorem is that *all bases of V have the same number of elements*. This number is called the *dimension* of V, written dim V (or sometimes $\dim_F V$, if it desired to mention explicitly the field F of scalars). The Basis Theorem is stated fully and proved below, but it is worthwhile to look first at some examples.

Example 251. The vector space $V = F^n$ has a basis $\{e_1, \ldots, e_n\}$ (see Example 250), hence dim $F^n = n$. The Basis Theorem tells us that any other basis of F^n will also have n elements. For example, the bases of R^2 found in Example 249 all have two elements.

Example 252. If $V = \{0\}$, then by the conventions explained at the end of the last section, the empty set $\varnothing$ is a basis for V. So dim $V = |\varnothing| = 0$ in this case. Any finitely generated vector space which has non-0 elements has dimension ≥ 1.

Example 253. The line $L(0, x)$ in R^n is, as we saw in Example 241, the subspace $\mathrm{sp}\{x\}$ spanned by the non-0 vector x. Since $x \neq 0$, the set $\{x\}$ is independent (Example 247), hence $\{x\}$ is a basis for $L(0, x)$. Therefore, as vector space over R, $L(0, x)$ has dimension 1. Similarly, the plane $L(0, x, y)$ is a vector space of dimension 2, since it is spanned by $\{x, y\}$, and (Example 247) $\{x, y\}$ must be a linearly independent set, hence it is a basis of $L(0, x, y)$. But in general, a line $L(a, b)$ or plane $L(a, b, c)$ are not vector spaces, so our definition of dimension does not apply to them. (Of course there is nothing to stop us extending the definition of dimension to sets like $L(a, b)$, $L(a, b, c)$, and this is done in text-books on geometry. See for example K. W. Gruenberg and A. J. Weir, *Linear Geometry*, Van Nostrand, London.)

Example 254. According to our definition, a vector space (such as $F[X]$, Example 241, section 9.5) which is not finitely generated, has no dimension. In fact it is possible to define bases even in these cases (with suitable modification of our definitions), but these bases are infinite sets. For example, the set $1, X, X^2, \ldots$ given in Example 244 is a basis for $F[X]$. In such a case, we say the space V is *infinite-dimensional*.

The Basis Theorem follows very easily from a lemma which is interesting in its own right, and is called the

EXCHANGE LEMMA. *Suppose that* $S = \{x_1, \ldots, x_p\}$ *is a subset of* V *which spans* V, *and that* $\{y_1, \ldots, y_m\}$ *is a linearly independent subset of* V. *Then* $m \leq p$, *and there is a subset* $\{x_{i_1}, \ldots, x_{i_{p-m}}\}$ *of* S *such that* $\mathrm{sp}\{y_1, \ldots, y_m, x_{i_1}, \ldots, x_{i_{p-m}}\} = V$. *In other words, we can 'exchange'* m *of the elements of* S *for the elements* $y_1, \ldots, y_m$, *and still have a set which spans* V.

Proof. Since $V = \mathrm{sp}\, S$, we can write $y_1 = \alpha_1 x_1 + \ldots + \alpha_p x_p$, for some $\alpha_1, \ldots, \alpha_p \in F$. Since $y_1 \neq \mathbf{0}$ (see Example 246), not all the α_i are zero. Suppose that $\alpha_j \notin 0$. We have $\alpha_j x_j = y_1 - \alpha_1 x_1 - \ldots - \alpha_{j-1} x_{j-1} - \alpha_{j+1} x_{j+1} - \ldots - \alpha_p x_p$, and then multiplying through by α_j^{-1} we find that

$$x_j \in \mathrm{sp}\, S(j),$$

where

$$S(j) = \{y_1, x_1, \ldots, x_{j-1}, x_{j+1}, \ldots, x_p\}. \tag{9.11}$$

Hence $V = \mathrm{sp}\, S(j)$, since any $v \in V$ can be written $v = \mu_1 x_1 + \ldots + \mu_j x_j + \ldots + \mu_p x_p$ for some $\mu_1, \ldots, \mu_p \in F$ (this is because $V = \mathrm{sp}\, S$), and we may now use (9.11) to replace the term $\mu_j x_j$ by a linear combination of $S(j)$, and we get an expression of v as a linear combination of $S(j)$.

We have now 'exchanged' x_j for y_1. Assuming $m \geq 2$, we next write y_2 as a linear combination of $S(j)$, say

$$y_2 = \lambda_1 y_1 + \beta_1 x_1 + \ldots + \beta_{j-1} x_{j-1} + \beta_{j+1} x_{j+1} + \ldots + \beta_p x_p. \tag{9.12}$$

We cannot have all the β_i's *equal to zero*, since that would leave us with $y_2 = \lambda_1 y_1$, i.e. $(-\lambda_1) y_1 + 1 y_2 = 0$, which is a non-trivial relation on $\{y_1, \ldots, y_m\}$. Let β_k be a non-zero element among $\beta_1, \ldots, \beta_{j-1}, \beta_{j+1}, \ldots, \beta_p$. Then we can use (9.12) to express x_k as a linear combination of the set $S(j, k) = \{y_1, y_2, \ldots, x_{j-1}, x_{j+1}, \ldots, x_{k-1}, x_{k+1}, \ldots, x_p\}$. Because $V = \mathrm{sp}\, S(j)$, this shows that $V = \mathrm{sp}\, S(j, k)$. Thus we have 'exchanged' x_j and x_k for y_1 and y_2. This process goes on as long as there are elements of $y_1, \ldots, y_m$ which have not been used in an exchange. At the end we have that $V = \mathrm{sp}\, \{y_1, \ldots, y_m x_{i_1}, \ldots, x_{i_{p-m}}\}$ where $x_{i_1}, \ldots, x_{i_{p-m}}$ are the 'unexchanged' elements of S. This proves in particular that $m \leq p$.

We come at last to our main theorem.

BASIS THEOREM. *Let V be a finitely generated vector space over the field F. Then*

(i) *Any two bases of V have the same number of elements. If this number is n, say that n is the dimension of V, and write*

   dim $V = n$.

(ii) *Any subset $S = \{x_1, \ldots, x_p\}$ of V which spans V has $p \geq n$ elements. If $p = n$, then S is a basis of V. If $p > n$, then by removing suitable elements of S, one can obtain a subset S' of S which is a basis of V.*

(iii) *Any linearly independent subset $I = \{y_1, \ldots, y_m\}$ of V has $m \leq n$ elements. If $m = n$, then I is a basis of V. If $m < n$, then by adding suitable elements $y_{m+1}, \ldots, y_n$ one can 'enlarge' I to a basis $\{y_1, \ldots, y_m, y_{m+1}, \ldots, y_n\}$ of V.*

Proof. (i) Let $\{x_1, \ldots, x_p\}$, $\{y_1, \ldots, y_m\}$ be two bases of V. Since $\{x_1, \ldots, x_p\}$ spans V and $\{y_1, \ldots, y_m\}$ is independent, the Exchange Lemma shows that $m \leq p$. But we can reverse the roles of these two sets, and apply the Exchange Lemma to show that $p \leq m$. Hence $m = p$, as required. From now on assume $m = p = n$, i.e.

   dim $V = n$.

(ii) From theorem (ii), section 9.6 we already know that there is a subset S' of S which is a basis of V. Since by (i) this basis S' has n elements, we have $n \leq p$. If $n = p$, we must have $S' = S$, i.e. in this case, S itself is a basis of V.

(iii) Apply the Exchange Lemma with $\{x_1, \ldots, x_n\}$ any basis of V, and $\{y_1, \ldots, y_m\}$ the given independent set I. This gives that $m \leq n$, and that $\{y_1, \ldots, y_m, x_{i_1}, \ldots, x_{i_{n-m}}\}$ spans V, for some elements $x_{i_1}, \ldots, x_{i_{n-m}}$ of X. But then this is a spanning set of n elements, so by (ii) it is a basis of V. In case $m = n$, this shows $\{y_1, \ldots, y_m\}$ is already a basis of V. If $m < n$, we can enlarge $\{y_1, \ldots, y_m\}$ to a basis of V by adding the elements $x_{i_1}, \ldots, x_{i_{n-m}}$.

Example 255. Dimension of a subspace. Suppose that U is a subspace of a vector space V of dimension n. Consider the set $\mathscr{B}$ of all finite subsets of U, which are linearly independent. At the very least, $\mathscr{B}$ contains the empty subset $\varnothing$ of U; if $U \neq \{\mathbf{0}\}$, $\mathscr{B}$ also contains all the one-element sets $\{u\}$, where u is a non-$\mathbf{0}$ element of U; however, the order m of any element

$\{u_1, \ldots, u_m\}$ of $\mathscr{B}$ is always $\leq n$, by (iii) of the Basis Theorem. So let m be the maximum order of elements of $\mathscr{B}$, and suppose that $\{u_1, \ldots, u_m\}$ is an element of $\mathscr{B}$ of this order. *Then* $\{u_1, \ldots, u_m\}$ *must span* U. For let u be any element of U. The set $\{u_1, \ldots, u_m, u\}$, since it has $m+1$ elements, must be linearly dependent, i.e. there is a non-trivial relation $\alpha_1 u_1 + \ldots + \alpha_m u_m + \alpha u = \mathbf{0}$, for some $\alpha_1, \ldots, \alpha_m, \alpha \in F$, not all zero. Now if α were zero, we would have a non-trivial relation on $\{u_1, \ldots, u_m\}$, which is impossible since this set is independent. So $\alpha \neq 0$, and $u = \alpha^{-1}\alpha_1 u_1 - \ldots - \alpha^{-1}\alpha_m u_m \in \mathrm{sp}\{u_1, \ldots, u_m\}$.

Thus $\{u_1, \ldots, u_m\}$, being linearly independent *and* a spanning set for U, is a basis of U, therefore dim $U = m \leq n$. So we have proved: *any subspace U of a finitely generated vector space V is itself finitely generated, and dim $U \leq$ dim V.*

9.8   Linear maps. Isomorphism of vector spaces

In this section we consider the kind of maps between vector spaces which are called *linear*; they could just as well be called 'homomorphisms of vector spaces' because, as we shall see, they are exactly analogous to homomorphisms of groups (Chapter 7) and to homomorphisms of rings (section 8.6).

DEFINITION. Let V, W be vector spaces over the same field F. A map $f: V \to W$ is called *linear* if it satisfies the following two conditions

VH1 $\qquad f(x+y) = f(x) + f(y)$

and

VH2 $\qquad f(\alpha f) = \alpha f(x)$,

for all x, $y \in V$ and all $\alpha \in F$.

Remarks

1   Condition **VH1** shows that a linear map $f: V \to W$ is a homomorphism between the additive groups $(V, +)$, $(W, +)$. So it follows from the lemmas of section 7.2 that $f(\mathbf{0}_V) = \mathbf{0}_W$ (here $\mathbf{0}_V$, $\mathbf{0}_W$ are the zero elements of V, W), and $f(-x) = -f(x)$ for all $x \in V$. (These facts could also be deduced from **VH2**, by taking $\alpha = 0_F$ and $\alpha = -1$, and using theorem (ii), (iii) of section 9.1.)

2 **VH1** and **VH2** together are equivalent to the single condition **VH**:$f(\alpha x + \beta y) = \alpha f(x) + \beta f(y)$, for all x, $y \in V$ and all α, $\beta \in F$. For **VH1** implies $f(\alpha x + \beta y) = f(\alpha x) + f(\beta y)$, and then **VH2** gives $f(\alpha x) = \alpha f(x), f(\beta x) = \beta f(x)$. So **VH1** and **VH2** together imply **VH**. Conversely if a map $f: V \to W$ satisfies **VH**, we can deduce **VH1** (by taking $\alpha = \beta = 1_F$) and **VH2** (by taking $\beta = 0_F$).

Isomorphisms, automorphisms, etc., for vector spaces

A linear map $f: V \to W$ is really a 'homomorphism of vector spaces', and for this reason the standard terms of group theory are often applied to linear maps of special kinds. In particular a linear map $f: V \to W$ which is bijective is called a *linear isomorphism*, or an *isomorphism* of vector spaces; one says that vector spaces V, W over F are *isomorphic* (and writes $V \cong W$) if there exists a linear isomorphism $f: V \to W$. A linear map $f: V \to V$ is called a *linear endomorphism of V* (or a *linear transformation of V*); if also f is bijective, it is called a *linear automorphism of V*.

We have the following lemmas, which correspond to lemmas (iv), (v) of section 7.2, and whose proofs we leave as an exercise for the reader.

LEMMA. (i) *If V, W, X are vector spaces over the the same field F, and if $f: V \to W$, $g: W \to X$ are both linear maps, then $gf: V \to X$ is a linear map. If f, g are both linear isomorphisms, then gf is a linear isomorphism.*
(ii) *If $f: V \to W$ is a linear isomorphism, then $f^{-1}: W \to V$ is a linear isomorphism.*

Kernel and image of a linear map

These are just like the kernel and image of a homomorphism of groups (section 7.4) or of rings (section 8.9).

DEFINITION. Let V, W be vector spaces, and let $f: V \to W$ be a linear map. Then the sets

$$\text{Ker } f = \{x \in V \,|\, f(x) = \mathbf{0}_W\} \tag{9.13}$$

and

$$\text{Im}\,f = \{f(x)\,|\,x \in V\} \tag{9.14}$$

are called the *kernel* and *image* of f, respectively.

We have then the

LEMMA. (iii) *Ker f is a subspace of V, and Im f is a subspace of W.*
(iv) *The linear map $f: V \to W$ is injective, if and only if Ker $f = \{\mathbf{0}_V\}$.*

Proof. (iii) If x, $y \in \text{Ker}\,f$, then by **VH1** $f(x+y) = f(x) + f(y) = \mathbf{0}_W + \mathbf{0}_W = \mathbf{0}_W$, so that $x + y \in \text{Ker}\,f$. If $\lambda \in F$, then by **VH2** $f(\lambda x) = \lambda f(x) = \lambda \mathbf{0}_W = \mathbf{0}_W$, so that $\lambda x \in \text{Ker}\,f$. This proves that Ker f satisfies the subspace conditions **SS1**, **SS2** (section 9.4). Next let a, b be elements of Im f. This implies there are elements x, y of V such that $f(x) = a$, $f(y) = b$. But then $a + b = f(x) + f(y) = f(x + y)$ belongs to Im f, as does $\lambda a = \lambda f(x) = f(\lambda x)$. So Im f is a subspace of W.
(iv) If f is injective, and if $x \in \text{Ker}\,f$, we have $f(x) = \mathbf{0}_W = f(\mathbf{0}_V)$, hence $x = \mathbf{0}_V$ (by injectivity of f). Hence Ker $f = \{\mathbf{0}_V\}$. Conversely suppose Ker $f = \{\mathbf{0}_V\}$. Let x, $y \in V$ be such that $f(x) = f(y)$. Then $f(x - y) = f(x) - f(y) = \mathbf{0}_W$, hence $x - y \in \text{Ker}\,f$, hence $x - y = \mathbf{0}_V$. Thus $f(x) = f(y)$ implies $x = y$, so that f is injective. This completes the proof of lemma (iv).

We shall see in section 9.12 that a 'Homomorphism Theorem' holds for vector spaces, just as for groups (section 7.4) and rings (section 8.9).

Example 256. Define $f: F^3 \to F^2$ as follows. If $(x, y, z) \in F^3$ let $f[(x, y, z)] = (2x + 3y + z, -x + y - z) \in F^2$. It is easy to check **VH1** and **VH2** and hence show that f is linear. We next prove that f is surjective (i.e. f is a *linear epimorphism*). For let (h, k) be an arbitrary element of F^2. Then $f[(x, y, z)] = (h, k)$ if and only if x, y, z satisfy the equations

$\text{E}(1) \qquad 2x + 3y + z = h$

and

$\text{E}(2) \qquad -x + y - z = k.$

We can solve these for x, y; first $\text{E}(1) + 2\text{E}(2)$ gives $5y - z = h + 2k$, while $\text{E}(1) - 3\text{E}(2)$ gives $5x + 4z = h - 3k$. From these we see that

$f[(x, y, z)] = (h, k)$, if and only if $x = \frac{1}{5}(h - 3k - 4z)$, $y = \frac{1}{5}(h + 2k + z)$. So, for example, f maps $[\frac{1}{5}(h - 3k), \frac{1}{5}(h + 2k), 0]$ to (h, k). This proves f is surjective. The kernel of f consists of all (x, y, z) such that $f[(x, y, z)] = (0, 0)$, and (taking $h = k = 0$ in the above) this happens if and only if $x = \frac{4}{5}z$, $y = \frac{1}{5}z$. Therefore $\operatorname{Ker} f = \{z(-\frac{4}{5}, \frac{1}{5}, 1) | z \in F\}$; $\operatorname{Ker} f$ is spanned by $(-\frac{4}{5}, \frac{1}{5}, 1)$, and this vector forms a basis of $\operatorname{Ker} f$.

Example 257. Let V be the vector space of all *polynomial functions* $f: R \to R$, i.e. of maps f which send $x \in R$ to $f(x) = a_0 + a_1 + \cdots + a_n x^n$ ($a_0, \ldots, a_n$ being fixed elements of R). (This polynomial function f is logically different from the *polynomial* $f(X) = a_0 + a_1 X + \cdots + a_n X^n \in R[X]$; for this reason we use small x in place of X, when dealing with the function.) Define map $D: V \to V$ by the rule: if $f = f(x) \in V$, let $Df = f'(x)$ (derived function). It is clear that D is linear, since from elementary calculus $D(f + g) = Df + Dg$ and $D(\alpha f) = \alpha D(f)$, for all f, $g \in V$ and $\alpha \in R$. The map D is surjective, because any $f \in V$ had form $f = Dg$ for some $g \in V$, for example take $g(x) = \int_0^x f(t) dt$. In fact we can define another map $J: V \to V$ by $J(f) = \int_0^x f(t) dt$ (so that, for example, $J(1 - x + 2x^2) = x - \frac{1}{2}x^2 + \frac{2}{3}x^3$). If we differentiate $J(f)$, we get f again; this shows $DJ = \iota_V$, the identity map of V. But $JD \neq \iota_V$, since $JD(f) = \int_0^x f'(t) dt = f(x) - f(0)$, which may not be the same as $f(x)$. The reader can check that $\operatorname{Ker} D$ is the set of all *constant* polynomial functions $f(x) = a_0$, $a_0 \in R$.

We end this section with one of the most important theorems on vector spaces. Suppose $x_1, \ldots, x_n$ are given elements of a vector space V. We define a map $f: F^n \to V$ as follows

$$f(\lambda) = f[(\lambda_1, \ldots, \lambda_n)] = \lambda_1 x_1 + \cdots + \lambda_n x_n$$

for all $\lambda = (\lambda_1, \ldots, \lambda_n) \in F^n$. This map is linear, because $f(\lambda + \mu) = (\lambda_1 + \mu_1)x_1 + \cdots + (\lambda_n + \mu_n)x_n = f(\lambda) + f(\mu)$, and $f(\alpha\lambda) = (\alpha\lambda_1)x_1 + \cdots + (\alpha\lambda_n)x_n = \alpha f(\lambda)$, for all λ, $\mu \in F^n$ and $\alpha \in F$. Clearly $\operatorname{Im} f$ consists of all linear combinations $\lambda_1 x_1 + \cdots + \lambda_n x_n (\lambda \in F^n)$, i.e. $\operatorname{Im} f = \operatorname{sp}\{x_1, \ldots, x_n\}$. Hence *the map* $f: F^n \to V$ *is surjective if and only if* $\{x_1, \ldots, x_n\}$ *spans* V. On the other hand f *is injective if and only if* $\{x_1, \ldots, x_n\}$ *is a linearly independent set.* For by lemma (iv) f is injective if and only if $\operatorname{Ker} f = \{\mathbf{0}\}$. But $\operatorname{Ker} f = \{\mathbf{0}\}$ if and only if $f(\lambda) = \mathbf{0} \Rightarrow \lambda = \mathbf{0}$, for all $\lambda \in F^n$, i.e. if and only if $\lambda_1 x_1 + \cdots + \lambda_n x_n \Rightarrow \lambda_1 = \cdots = \lambda_n = 0$, which is the condition for $\{x_1, \ldots, x_n\}$ to be independent.

Putting these two facts together, we see that *if* $\{x_1, \ldots, x_n\}$ *is a basis of V, then* $f: F^n \to V$ *is bijective, i.e. f is a linear isomorphism.* This proves the following theorem.

THEOREM. *If V is a vector space over F of dimension n, then* $V \cong F^n$. *In fact if* $\{x_1, \ldots, x_n\}$ *is a basis for V, then the map* $f: F^n \to V$ *defined above is a linear isomorphism*

This means there are very few different 'types' of finitely generated vector space V over a given field! Every vector space V of dimension n is isomorphic to the vector space F^n.

9.9 Matrices

Further study of linear maps requires the use of *matrices*. So we interrupt our account of vector-space theory to discuss these.

Let m, n be positive integers, and F a field. Then an $m \times n$ *matrix over F* is a rectangular array

$$A = \begin{bmatrix} a_{11}, & a_{12}, & \ldots, & a_{1n} \\ a_{21}, & a_{22}, & \ldots, & a_{2n} \\ \vdots & \vdots & \vdots & \vdots \\ a_{m1}, & a_{m2}, & \ldots, & a_{mn} \end{bmatrix} \qquad (9.15)$$

where the a_{ij} are elements of F. This matrix is often presented in one of the forms

$$A = (a_{ij}), \qquad \text{or } A = (a_{ij})_{i=1,\ldots,m; j=1,\ldots,n}.$$

The element a_{ij} is the (i, j)-*entry* or (i, j)-*component* of A. The pair (m, n) is called the *shape* of A. A matrix of shape (n, n) is called a *square matrix*.

Matrices $A = (a_{ij})$ and $B = (b_{ij})$ are said to be *equal* (and we write $A = B$) if and only if they both have the same shape (m, n), and $a_{ij} = b_{ij}$ for all $i = 1, \ldots, m$ and all $j = 1, \ldots, n$.

Matrices can be thought of as a generalization of vectors. In fact a matrix of shape $(1, n)$ is the same as an n-vector, as defined in section 9.2. A matrix of shape $(m, 1)$ looks like

$$\begin{bmatrix} u_1 \\ u_2 \\ \vdots \\ u_m \end{bmatrix},$$

and is called a *column m-vector* (an ordinary *n*-vector is often called a *row n-vector*). If $A = (a_{ij})$ is a matrix of arbitrary shape (m, n) we refer to the *n*-vectors

$$(a_{11}, a_{12}, \ldots, a_{1n}), (a_{21}, a_{22}, \ldots, a_{2n}), \ldots, (a_{m1}, a_{m2}, \ldots, a_{mn})$$

as the *rows of A*, and to the column *m*-vectors

$$\begin{bmatrix} a_{11} \\ a_{21} \\ \vdots \\ a_{m1} \end{bmatrix}, \begin{bmatrix} a_{12} \\ a_{22} \\ \vdots \\ a_{m2} \end{bmatrix}, \ldots, \begin{bmatrix} a_{1n} \\ a_{2n} \\ \vdots \\ a_{mn} \end{bmatrix}$$

as the *columns of A*.

Addition of matrices. Multiplication by scalars

If $A = (a_{ij})$, $B = (b_{ij})$ are matrices of the same shape (m, n), we define their *sum* to be the $m \times n$ matrix $A + B$ whose (i, j)-entry is $a_{ij} + b_{ij}$. If α is any scalar (i.e. α is any element of F) we define αA to be the $m \times n$ matrix whose (i, j)-entry is αa_{ij}. The zero matrix of shape (m, n), denoted $0_{m,n}$ (or simply 0 if it is clear what shape is intended), is the $m \times n$ matrix whose (i, j)-entry is 0 (this being the zero 0_F of F), for all $i = 1, \ldots, m; j = 1, \ldots, n$.

Example 258.

$$A = \begin{bmatrix} 2 & 1 \\ -7 & 1 \\ 0 & -2 \end{bmatrix} \quad \text{and} \quad B = \begin{bmatrix} -4 & -1 \\ 0 & 6 \\ 2 & 1 \end{bmatrix}$$

are both 3×2 matrices (i.e. matrices of shape $(3, 2)$) over R. We have

$$A + B = \begin{bmatrix} 2+(-4) & 1+(-1) \\ (-7)+0 & 1+6 \\ 0+2 & (-2)+1 \end{bmatrix} = \begin{bmatrix} -2 & 0 \\ -7 & 7 \\ 2 & -1 \end{bmatrix},$$

and

$$8B = \begin{bmatrix} -32 & -8 \\ 0 & 48 \\ 16 & 8 \end{bmatrix}.$$

The matrix $(-1)A$ is usually denoted $-A$, so that in our case

$$-A = \begin{bmatrix} -2 & -1 \\ 7 & -1 \\ 0 & 2 \end{bmatrix}.$$

Note

$$A + (-A) = \begin{bmatrix} 0 & 0 \\ 0 & 0 \\ 0 & 0 \end{bmatrix} = 0_{3,2}.$$

THEOREM. *Let F be a field, and m, n any positive integers. Then the set $F^{m,n}$ of all $m \times n$ matrices over F becomes a vector space over F, when we define the sum $A + B$ and scalar action αA as above. The zero element of $F^{m,n}$ is the zero matrix $0_{m,n}$. The negative of a matrix A is the matrix $(-1)A = -A = (-a_{ij})$.*

Proof. This consists in checking the vector space axioms **FV1–FV8** for $V = F^{m,n}$. This is straightforward, and is left to the reader. We call attention to the

NOTATION. The set of all $m \times n$ matrices over F is denoted $F^{m,n}$. In particular $F^{1,n}$ is the same as the set F^n of all n-vectors over F, and $F^{m,1}$ is the set of all column m-vectors over F.

Example 259. For each $i \in \{1, \ldots, m\}$ and $j \in \{1, \ldots, n\}$, let E_{ij} be the $m \times n$ matrix whose (i, j)-entry is 1, and whose other entries are all zero. Then *the*

set $S = \{E_{ij} | i = 1, \ldots, m; j = 1, \ldots, n\}$ is a basis of the vector space $F^{m,n}$. To prove this, we verify the basis conditions **B1**, **B2** (p. 182).

B1 S spans $F^{m,n}$, since an arbitrary matrix $A = (a_{ij}) \in F^{m,n}$ can be written $A = \sum a_{ij} E_{ij}$ (the sum is over all $i = 1, \ldots, m$ and all $j = 1, \ldots, n$), i.e. as a linear combination of the E_{ij}.

B2 S is linearly independent, since if we have any linear relation $\sum a_{ij} E_{ij} = 0_{m,n}$, this is the same as saying that the matrix $A = (a_{ij})$ is equal to $0_{m,n}$, i.e. that $a_{ij} = 0$ for all i, j. So the only linear relation on S is the trivial one.

It is clear that the number of elements of S is mn, hence *the dimension of the vector space $F^{m,n}$ is mn.*

Multiplication of matrices

The sum $A + B$ of matrices A, B, as we have just seen, is defined only when A, B have the same shape, and is given by the simple rule: the (i, j)-entry of $A + B$ is the sum of the (i, j)-entries of A and B. The definition of the *product* AB of matrices A, B is very different.

DEFINITION. The product AB of matrices $A = (a_{ij})$, $B = (b_{hk})$ is defined if and only if A has shape (m, n) and B has shape (n, p), i.e. if and only if

Number of columns of A = number of rows of B.   (9.16)

Suppose that (9.16) holds, then $AB = C$ is the matrix of shape (m, p), whose (i, k)-entry is given by the rule

$$c_{ik} = a_{i1} b_{1k} + \cdots + a_{in} b_{nk} = \sum_{j=1}^{n} a_{ij} b_{jk}, \qquad (9.17)$$

for all $i = 1, \ldots, m$ and all $k = 1, \ldots, p$.

To get some understanding of this rule, first take the case $m = 1$, $p = 1$. Then A is a row n-vector, and B is a column n-vector. The definition above says that AB is a matrix of shape $(1, 1)$, i.e. AB is a 1×1 matrix (c_{11}), where (according to (9.17)) the element $c_{11} \in F$ is given by

$$c_{11} = a_{11} b_{11} + \cdots + a_{1n} b_{n1}.$$

A 1×1 matrix (c_{11}) is usually regarded simply as a scalar, i.e. we write (c_{11}) as c_{11}. Let us change our notation for A, B by writing $a_{1j}=u_j$, $b_{j1}=v_j$, for $j=1,\ldots,n$. Then the formula for the product of the matrices A, B becomes

$$(u_1,\ldots,u_n)\begin{bmatrix} v_1 \\ \vdots \\ v_n \end{bmatrix} = u_1v_1 + \cdots + u_nv_n. \qquad (9.18)$$

This particular product of matrices is often called the *scalar product*; it 'multiplies' the row n-vector u, by the column n-vector v, to give the scalar $uv=u_1v_1 + \cdots + u_nv_n$. (Sometimes this product is written $u.v$.)

Now go back to our general case, where A, B have shapes (m, n), (n, p) respectively. Formula (9.17) says that c_{ik} is just the scalar product of the ith row of A, with the kth column of B. This is the easiest way to remember the rule for matrix multiplication:

$$(i, k)\text{-entry of } AB = (i\text{th row of } A)(k\text{th column of } B). \qquad (9.19)$$

This also shows why condition (9.16) is necessary; it is necessary because the scalar product (9.18) of a row vector with a column vector is defined only if these vectors have the same length n.

Example 260.

$$\begin{bmatrix} 1 & 0 & -1 & 2 \\ 3 & -3 & 0 & 1 \end{bmatrix} \begin{bmatrix} 1 & 1 & -1 \\ 6 & -2 & 0 \\ -4 & 0 & 1 \\ 1 & 1 & 1 \end{bmatrix} = \begin{bmatrix} 7 & 3 & 0 \\ -14 & 10 & -2 \end{bmatrix}.$$

If A, B are the 2×4, 4×3 matrices on the left, they satisfy condition (9.16), so AB is defined and is a 2×3 matrix C. Use (9.19) to calculate the entries in $AB=C$. For example, the (2, 3)-entry of C is (row 2 of A) (col 3 of B)$=3.(-1)+(-3).0+0.1+1.1=-2$. Notice BA is not defined in this case.

Example 261. If $A = (a_{ij})$ has shape (m, n), and if $0_{n,p}$ is the zero matrix of shape (n, p), then $A0_{n,p}=0_{m,p}$. This is because each column of $0_{n,p}$ is the zero vector, and so the product of ith row of A with any column of $0_{n,p}$ is zero. Similarly we have $0_{m,n}B=0_{m,p}$, for any B of shape (n, p).

Example 262. The product of a column vector v with a row vector u of the same length, is defined since we regard $v = A$ as a $n \times 1$ matrix, and $u = B$ as $1 \times n$ matrix, so condition (9.16) is satisfied; the answer is an $n \times n$ matrix. For example

$$\begin{bmatrix} v_1 \\ v_2 \\ v_3 \end{bmatrix} \begin{bmatrix} u_1 & u_2 & u_3 \end{bmatrix} = \begin{bmatrix} v_1 u_1 & v_1 u_2 & v_1 u_3 \\ v_2 u_1 & v_2 u_2 & v_2 u_3 \\ v_3 u_1 & v_3 u_2 & v_3 u_3 \end{bmatrix}$$

9.10 Laws of matrix algebra. The ring $M_n(F)$

Matrices are used in many parts of mathematics, and we shall apply them to the theory of vector spaces. But before we come to these applications, it is worth spending a little more time on the formal properties of the addition, scalar multiplication, and multiplication of matrices, i.e. on 'matrix algebra'. We have already shown (theorem, section 9.9) that the set $F^{m,n}$ of all $m \times n$ matrices over F is a vector space. The next theorem gives the main properties of matrix multiplication.

THEOREM. *Let m, n, p, q be positive integers. Then*

(i) $A(B+C) = AB + AC$,   *for any $A \in F^{m,n}$, $B \in F^{n,p}$, $C \in F^{n,p}$.*

(ii) $(A+B)C = AC + BC$,   *for any $A \in F^{m,n}$, $B \in F^{m,n}$, $C \in F^{n,p}$.*

(iii) $\lambda(AB) = (\lambda A)B = A(\lambda B)$,   *for any $A \in F^{m,n}$, $B \in F^{n,p}$, $\lambda \in F$.*

(iv) $(AB)C = A(BC)$,   *for any $A \in F^{m,n}$, $B \in F^{n,p}$, $C \in F^{p,q}$.*

Proof. Suppose that a_{ij}, b_{ij}, c_{ij} denote, respectively, the (i, j)-entries of the matrices A, B, C.

(i) First notice that both sides of this equation are of the same shape (m, p). Take any $i \in \{1, \ldots, m\}$ and $k \in \{1, \ldots, p\}$. The rules for matrix addition and multiplication give us that the (i, k)-entry of $A(B+C)$ is $a_{i1}(b_{1k} + c_{1k}) + \cdots + a_{in}(b_{nk} + c_{nk})$. The (i, k)-entry of $AB + AC$, on the other hand, is $(a_{i1}b_{1k} + \cdots + a_{in}b_{nk}) + (a_{i1}c_{1k} + \cdots + a_{in}c_{nk})$. It is clear that these entries are equal (we use the fact that F is a ring to prove this, of course); then since $A(B+C)$ and

$AB + AC$ have the same (i, k)-entries, for all i, k, they are equal. The proofs of (ii), (iii) proceed on the same lines, and are left to the reader. The proof of (iv), the important *associative law for matrix multiplication*, goes as follows: first notice that both $(AB)C$ and $A(BC)$ are of shape (m, q). Take any $i \in \{1, \ldots, m\}$ and $l \in \{1, \ldots, q\}$. Let $AB = U$. U has (i, k)-entry $u_{ik} = \sum_{h=1}^{n} a_{ih}b_{hk}$, for any $k \in \{1, \ldots, p\}$. So the (i, l)-entry of $(AB)C = UC$ is

$$\sum_{k=1}^{p} u_{ik}c_{kl} = \sum_{k=1}^{p} \left[\sum_{h=1}^{n} (a_{ih}b_{hk})c_{kl} \right]. \tag{9.20}$$

Now let $BC = V$. V has (h, l)-entry $v_{hl} = \sum_{k=1}^{p} b_{hk}c_{kl}$, for any $h \in \{1, \ldots, n\}$. So the (i, l)-entry of $A(BC) = AV$ is

$$\sum_{h=1}^{n} a_{ih}v_{hl} = \sum_{h=1}^{n} \left[\sum_{k=1}^{p} a_{ih}(b_{hk}c_{kl}) \right]. \tag{9.21}$$

Now $(a_{ih}b_{hk})c_{kl} = a_{ih}(b_{hk}c_{kl})$ by the fact that multiplication on F is associative; we shall write $f(h, k) = a_{ih}b_{hk}c_{kl}$ for short. Hence (9.20) and (9.21) are equal, because they are, respectively, the left and right sides of the equation

$$\sum_{k=1}^{p} \sum_{h=1}^{n} f(h, k) = \sum_{h=1}^{n} \sum_{k=1}^{p} f(h, k), \tag{9.22}$$

which holds because each side is simply the sum of all the np terms in the array

$$\begin{array}{cccc} f(1, 1) & f(1, 2) & \ldots & f(1, p) \\ f(2, 1) & f(2, 2) & \ldots & f(2, p) \\ \vdots & & & \vdots \\ f(n, 1) & f(n, 2) & \ldots & f(n, p) \end{array} \tag{9.23}$$

the left side of equation (9.22) being got by first adding up the terms in each column of (9.23), and then adding up these 'column sums', while to get the right side, you first add up the terms in each row of (9.23), and then add up these 'row sums'. Since addition in F is both associative and commutative, we get the same result in each case.

(Equation (9.22) is called the *double summation formula*, and is useful in many other situations.)

Identity matrix

Let n be a positive integer. Then the $n \times n$ matrix

$$I_n = \begin{bmatrix} 1 & 0 & \dots & 0 \\ 0 & 1 & \dots & 0 \\ \vdots & \vdots & & \vdots \\ 0 & 0 & \dots & 1 \end{bmatrix},$$

whose (i, j)-entry is $1 (= 1_F)$ *if* $i = j$, and is $0 (= 0_F)$ if $i \neq j$, is called the $n \times n$ *identity matrix*. It is very easy to prove

THEOREM (v) $I_m A = A = A I_n$, *for all* $A \in F^{m,n}$.

Example 263. Scalar matrices. If $\lambda \in F$, the $n \times n$ matrix λI_n is sometimes called a *scalar matrix*, because multiplication by an arbitrary $m \times n$ matrix A by λI_n produces the same effect as the multiplication of A by the scalar λ; we have in fact $(\lambda I_m) A = \lambda A = A(\lambda I_n)$. λI_n has all its 'diagonal' entries equal to λ, while the 'non-diagonal' entries are all zero (the (i, i)-entries of a 'square' (i.e. $n \times n$) matrix are often called the 'diagonal entries'):

$$\lambda I_m = \begin{bmatrix} \lambda & 0 & \dots & 0 \\ 0 & \lambda & \dots & 0 \\ \vdots & \vdots & \vdots & \vdots \\ 0 & 0 & \dots & \lambda \end{bmatrix}.$$

Transpose

If $A = (a_{ij})$ is an $m \times n$ matrix, then the *transpose of A*, is the $n \times m$ matrix A^T whose (i, j)-entry is equal to the (j, i)-entry of A, for all $i = 1, \dots, n$ and $j = 1, \dots, m$.† This can be said as follows: to get

---

†Other common notations for A^T are A^t and A'.

column i of A^T, take row i of A and write it as a column (and, similarly, the rows of A^T are the columns of A); for example

$$A = \begin{bmatrix} a_1 & a_2 \\ b_1 & b_2 \\ c_1 & c_2 \end{bmatrix}$$

has transpose

$$A^T = \begin{bmatrix} a_1 & b_1 & c_1 \\ a_2 & b_2 & c_2 \end{bmatrix}.$$

The transpose is useful in dealing with row and column vectors, because if $u = (u_1, \ldots, u_n)$ is a row n-vector, then

$$u^T = \begin{bmatrix} u_1 \\ u_2 \\ \vdots \\ u_n \end{bmatrix}$$

is the corresponding column n-vector; conversely if v is a column n-vector then v^T is the corresponding row n-vector. The main properties of the transpose are given in the next theorem.

THEOREM. (vi) $(A + B)^T = A^T + B^T$,   for all $A \in F^{m,n}$, $B \in F^{m,n}$.

(vii) $(\lambda A)^T = \lambda A^T$,   for all $A \in F^{m,n}$, $\lambda \in F$.

(viii) $(A^T)^T = A$,   for all $A \in F^{m,n}$.

(ix) $(AB)^T = B^T A^T$,   for all $A \in F^{m,n}$, $B \in F^{m,p}$.

Proof. The first three are proved easily, and these proofs are left to the reader. To prove (ix), write $C = AB$, so that, for any $i \in \{1, \ldots, m\}$, $k \in \{1, \ldots, p\}$, the (i, k)-entry of C is $c_{ik} = a_{i1}b_{1k} + \cdots + a_{in}b_{nk}$. Let $A^T = P$, $B^T = Q$, so that P and Q are, respectively, $n \times m$ and $p \times n$ matrices. The (j, i)-entry of P is $p_{ji} = a_{ij}$, and the (k, j)-entry of Q is $q_{kj} = b_{jk}$, for all $j \in \{1, \ldots, n\}$. Therefore the (k, j)-entry of QP is $q_{k1}p_{1i} + \cdots + q_{kn}p_{ni} = b_{1k}a_{i1} + \cdots + b_{nk}a_{in} = a_{i1}b_{1k} + \cdots + a_{in}b_{nk} = c_{ik}$. So the (k, i)-entry of

$B^T A^T = QP$ is the same as the (k, i)-entry of $C^T = (AB)^T$, and this proves (ix). Notice that we used the fact that multiplication in F is commutative.

Example 264. We can check the formula $(AB)^T = B^T A^T$, taking the matrices A, B of Example 260. We have

$$A^T = \begin{bmatrix} 1 & 3 \\ 0 & -3 \\ -1 & 0 \\ 2 & 1 \end{bmatrix}, \quad B^T = \begin{bmatrix} 1 & 6 & -4 & 1 \\ 1 & -2 & 0 & 1 \\ -1 & 0 & 1 & 1 \end{bmatrix},$$

$$B^T A^T = \begin{bmatrix} 7 & -14 \\ 3 & 10 \\ 0 & -2 \end{bmatrix}.$$

Example 265. Suppose $u = (u_1, \ldots, u_n)$, $v = (v_1, \ldots, v_n)$ are both row n-vectors. Then $uv^T = u_1 v_1 + \cdots + u_n v_n$, which is just another version of the 'scalar product' formula (9.18).

The matrix ring $M_n(F)$

Since the product of two $n \times n$ matrices is again an $n \times n$ matrix, the set $F^{n,n}$ is closed to the operation of matrix multiplication. In fact $F^{n,n}$ *is a ring*, as we see by verifying the ring axioms (section 8.1). Axioms **R1**–**R4** are satisfied, i.e. $F^{n,n}$ is an Abelian group with respect to addition, as we know by the fact that $F^{n,n}$ is a vector space over F (section 9.9, theorem). Axioms **R5** and **R6** hold, by theorem (iv), (v) of this section; the identity element of $F^{n,n}$ is the identity matrix I_n. Finally **R5** and **R6** follow from theorems (i) and (ii) of this section.

DEFINITION. The set $F^{n,n}$ of all $n \times n$ matrices over F, regarded as a ring, is called the *matrix ring $M_n(F)$*.

If $n = 1$, then $M_n(F)$ is the set of all 1×1 matrices (a_{11}), which we identify with the scalars a_{11}, so that $M_1(F) = F$. If $n \geq 2$, then $M_n(F)$ is not commutative. For example, take

$$A = \begin{bmatrix} 0 & 1 \\ 0 & 0 \end{bmatrix}, \qquad B = \begin{bmatrix} 1 & 0 \\ 0 & 0 \end{bmatrix};$$

then

$$AB = \begin{bmatrix} 0 & 0 \\ 0 & 0 \end{bmatrix}, \qquad BA = \begin{bmatrix} 0 & 1 \\ 0 & 0 \end{bmatrix},$$

so that $AB \neq BA$; hence $M_2(F)$ is a non-commutative ring. The reader may like to find such non-commuting pairs of $n \times n$ matrices A, B, for arbitrary $n \geq 2$.

Example 266. For given n, define map $f: F \to M_n(F)$ by the rule: $f(\lambda) = \lambda I_n$, for all $\lambda \in F$. Thus f maps each scalar $\lambda \in F$, to the *scalar matrix* λI_n (Example 263). It is easy to check that $f(1) = I_n$, $f(\lambda + \mu) = f(\lambda) + f(\mu)$, and $f(\lambda\mu) = f(\lambda)f(\mu)$, for all λ, $\mu \in F$. Thus f *is a homomorphism of rings* (section 8.6). The kernel of f is $\{0\}$, since $f(\lambda) = 0$ only if $\lambda = 0$. Thus f is injective (Example 202, section 8.9; this could be proved directly, of course), hence f induces an isomorphism $\bar{f}$ of F onto $\text{Im} f = \{\lambda I | \lambda \in F\}$ (by the Homomorphism Theorem for rings, section 8.9). In fact $\bar{f}$ is just the original map f, but regarded as map $F \to \text{Im} f$. So F is 'represented' by the subring $\{\lambda I | \lambda \in F\}$ of $M_n(F)$.

Example 267. Representation of the complex field by 2×2 real matrices. Let I, P be the elements of $M_2(R)$ given by

$$I = \begin{bmatrix} 1 & 0 \\ 0 & 1 \end{bmatrix}, \qquad P = \begin{bmatrix} 0 & 1 \\ -1 & 0 \end{bmatrix}.$$

It is clear that $P^2 = -I$. So we have a real matrix P, which 'copies' the defining property of the complex number i, namely $i^2 = -1$. Define a map $f: C \to M_2(R)$ as follows

$$f(a + ib) = aI + bP = \begin{bmatrix} a & b \\ -b & a \end{bmatrix},$$

for all a, $b \in R$. It is now easy to check that f is a homomorphism of rings, and that f is injective. So f induces an isomorphism $\bar{f}$ of C onto the subring S of $M_2(R)$ which consists of all matrices

$$\begin{bmatrix} a & b \\ -b & a \end{bmatrix}, \qquad a, b \in R.$$

Example 268. Representation of the quaternion division ring by 2×2 *complex matrices.* Let I, P, Q, R be the elements of $M_2(C)$

$$I = \begin{bmatrix} 1 & 0 \\ 0 & 1 \end{bmatrix}, \qquad P = \begin{bmatrix} 0 & 1 \\ -1 & 0 \end{bmatrix}, \qquad Q = \begin{bmatrix} i & 0 \\ 0 & -i \end{bmatrix},$$

$$R = \begin{bmatrix} 0 & -i \\ -i & 0 \end{bmatrix}.$$

We use matrix multiplication to check the following:

$$P^2 = Q^2 = R^2 = -I, \qquad PQ = R = -QP, \qquad QR = P = -RQ,$$
$$RP = Q = -PR.$$

But this shows that P, Q, R obey the same *multiplication rules* as the symbols i, j, k in the quaternion ring H (see Example 179, section 8.4). So if we define a map $f: H \rightarrow M_2(C)$ as follows

$$f(a_0 + a_1 i + a_2 j + a_3 k) = a_0 I + a_1 P + a_2 Q + a_3 R$$

$$= \begin{bmatrix} a_0 + ia_2 & a_1 - ia_3 \\ -a_1 - ia_3 & a_0 - ia_2 \end{bmatrix},$$

then f is a homomorphism of rings. (We should be careful to avoid confusing the symbol $i \in H$, with the symbol $i \in C$! This ought not to give trouble, since the entries of all the matrices must be elements of C.) It is also clear that f is injective, and so it induces an isomorphism $\bar{f}$ of H onto the subring S of $M_2(C)$ which consists of all matrices of the form

$$\begin{bmatrix} a_0 + ia_2 & a_1 - ia_3 \\ -a_1 - ia_3 & a_0 - ia_2 \end{bmatrix},$$

with $a_0, a_1, a_2, a_3 \in R$. We express this fact by saying that we have a 'representation' of the quaternion ring H by 2×2 complex matrices.

This representation provides an easy way to check that quaternion multiplication is associative. Let x, y, z be any elements of H. Then $f[x(yz)] = f(x)f(yz) = f(x)[f(y)f(z)]$, using the fact that f is a multiplicative homomorphism. Similarly $f[(xy)z] = [f(x)f(y)]f(z)$. But $f(x)$, $f(y)$, $f(z)$ are *matrices*, and we know that matrix multiplication is associative (theorem (iv) of this section). So $f(x)[f(y)f(z)] = [f(x)f(y)]f(z)$, i.e. $f[x(yz)] = f[(xy)z]$. But f is injective, hence $x(yz) = (xy)z$, which is what we wanted to prove.

9.11 Row space of a matrix. Echelon matrices

Suppose we are given an $m \times n$ matrix over F,

$$A = \begin{bmatrix} a_{11} & a_{12} & \cdots & a_{1n} \\ a_{21} & a_{22} & \cdots & a_{2n} \\ a_{m1} & a_{m2} & \cdots & a_{mn} \end{bmatrix}.$$

Each row $a_i = (a_{i1}, a_{i2}, \ldots, a_{in})$ is a row n-vector, i.e. an element of the vector space $F^n = F^{1,n}$. The subspace of F^n spanned by the m rows $a_1, \ldots, a_m$ of A is called the *row space of A*, and denoted row sp A. Its dimension is called the *row rank of A*. Similarly we define the *column space* col sp A to be the subspace of $F^{m,1}$ spanned by the n columns of A; the dimension of col sp A is called the *column rank* of A. We shall find later (section 9.13) that the row rank and column rank are equal (and so this number is just called the *rank* of A); for the moment, we shall concentrate on the problem: *Given a matrix $A \in F^{m,n}$, to find its row rank*. Since the row rank of A is the dimension of the row space of A, and since the dimension of a vector space is, by definition, the number of elements in any basis of the vector space, we are led to the more precise problem: *Given a matrix $A \in F^{m,n}$, to find a basis for row space of A*. In this section we shall give a method to solve this problem. The method (which is well adapted to machine calculation) can also be applied to the solution of *systems of linear equations*: we discuss these in section 9.12.

Example 269. The rows of the identity matrix I_n are just the unit vectors $e_1 = (1, 0, \ldots, 0), \ldots, e_n = (0, 0, \ldots, 1)$, and we know that these span the whole of F^n, and in fact form a basis of F^n. So row sp $I_n = F^n$, and row rank $I_n = \dim F^n = n$.

Elementary row operations

Let A and B be $m \times n$ matrices over F, with rows $a_1, \ldots, a_m$ and $b_1, \ldots, b_m$, respectively. We say A and B are *row equivalent* (and write $A \underset{r}{\sim} B$) if A and B have the same row space, i.e. if

$$\mathrm{sp}\{a_1, \ldots, a_m\} = \mathrm{sp}\{b_1, \ldots, b_m\}. \tag{9.24}$$

We shall describe certain *elementary row operations*, and show that each of these, when applied to a given matrix A, gives a new matrix B which is row equivalent to A. In what follows i, j, h stand for elements of $\{1, \ldots, m\}$ with $i \neq j$. The elementary row operations are of two types I, II.

Type I

Operation $R_i \leftrightarrow R_j$ simply interchanges the ith row of A with its jth row. The resulting matrix B has the same rows as A, although in a different order; hence (9.24) holds.

Type II

Operation $R_i \rightarrow \lambda R_i + \mu R_j$ leaves all rows of A unchanged except the ith, which is replaced by $\lambda(i\text{th row}) + \mu(j\text{th row})$; here λ, μ are any elements of F such that $\lambda \neq 0$. So the rows of the new matrix B are as follows

$$b_h = a_h \text{ (all } h \neq i), \qquad b_i = \lambda a_i + \mu a_j. \tag{9.25}$$

This shows that every row of B is a linear combination of the rows of A, hence $\mathrm{sp}\{b_1, \ldots, b_m\} \subseteq \mathrm{sp}\{a_1, \ldots, a_m\}$. But because $\lambda \neq 0$, we can use (9.25) to express each row of A as a linear combination of the rows of B:

$$a_h = b_h \text{ (all } h \neq i), \qquad a_i = \lambda^{-1}(b_i - \mu a_j) = \lambda^{-1} b_i - \lambda^{-1} \mu b_j \tag{9.26}$$

(remember that $i \neq j$, so $b_j = a_j$). Therefore $\mathrm{sp}\{a_1, \ldots, a_m\} \subseteq \mathrm{sp}\{b_1, \ldots, b_m\}$; hence (9.24) holds, and $A \underset{r}{\sim} B$.

Remark

It is quite common to break up an operation of type II into two simpler operations $R_i \rightarrow \lambda R_i$ ($\lambda \neq 0$), and $R_i \rightarrow R_i + \mu R_j$. Of course these are the special cases of our operation $R_i \rightarrow \lambda R_i + \mu R_j$, which you get by taking $\mu = 0$, and $\lambda = 1$, respectively.

Echelon matrices

Our next objective is to prove that any given $m \times n$ matrix A can be changed, by a sequence of elementary row operations, into an *echelon matrix B*. To understand the definition of an echelon (or *stepped*) matrix, it is a good idea to have an example to look at:

$$B = \begin{bmatrix} 0 & 0 & 1 & -2 & 0 & -4 & 7 \\ 0 & 0 & 0 & 0 & 3 & 5 & 0 \\ 0 & 0 & 0 & 0 & 0 & -8 & 2 \\ 0 & 0 & 0 & 0 & 0 & 0 & 0 \end{bmatrix}$$

DEFINITION. An $m \times n$ matrix $B = (b_{ij})$ is an echelon matrix if it satisfies the following two conditions:

Ech1. There is an integer r $(0 \le r \le m)$ such that the first r rows $b_1, \ldots, b_r$ are all non-$\mathbf{0}$, while the remaining rows (if any) are all equal to $\mathbf{0}$.

Ech2. For each of the non-$\mathbf{0}$ rows b_i let $j(i)$ be the smallest $j \in \{1, \ldots, n\}$ such that $b_{ij} \ne 0$. Then we require that

$$j(1) < j(2) < \cdots < j(r).$$

Our example above is a 4×7 matrix B which satisfies these conditions. **Ech1** holds with $r = 3$ – this condition says that the zero rows of B (if there are any) must all lie below the non-zero rows. **Ech2**, in our example, holds, since we see that $j(1) = 3$, $j(2) = 5$, $j(3) = 6$. This means that the non-zero entries in B all lie above the 'stepped' line indicated.

LEMMA. *If B is an $m \times n$ matrix satisfying conditions* **Ech1**, **Ech2** *above, then* $\{b_1, b_2, \ldots, b_r\}$ *is a linearly independent set. Hence for an echelon matrix B, the non-$\mathbf{0}$ rows of B form a basis of row sp B, and the rank of B is the number of non-zero rows of B.*

Proof. Let $\lambda_1 b_1 + \cdots + \lambda_r b_r = \mathbf{0}$ be any linear relation, with $\lambda_1, \ldots, \lambda_r \in F$. Since the only non-zero entry in the $j(1)$th column of B is $b_{1,j(1)}$ (see the example above), the $j(1)$th component of the

n-vector $\lambda_1 b_1 + \cdots + \lambda_n b_n$ is $\lambda_1 b_{1,j(1)}$. But this is zero, hence $\lambda_1 = (\lambda_1 b_{1,j(1)}) b_{1,j(1)}^{-1} = 0$. Our relation now reads $\lambda_2 b_2 + \cdots + \lambda_r b_r = \mathbf{0}$; we can ignore the row b_1 from now on. By an argument like that above, we find that $\lambda_2 = 0$. Eventually we have $\lambda_1 = \lambda_2 = \cdots = \lambda_r = 0$.

This proves that $\{b_1, \ldots, b_r\}$ is an independent set. But it also spans row sp B (since the other rows are zero, and so do not contribute anything (except $\mathbf{0}$) to sp$\{b_1, \ldots, b_n\}$). Therefore $\{b_1, \ldots, b_r\}$ is a basis of row sp B, and this shows that rank $B = \dim$ (row sp B) $= r$. The lemma is proved.

Our solution of the problem: given a matrix $A \in F^{m,n}$, to find a basis of row space of A, now rests on the following theorem.

THEOREM. *Given $A \in F^{m,n}$, it is possible by a finite sequence of elementary row operations to change A into an echelon matrix B.*

For each elementary operation leaves the row space unchanged, so that our final matrix B has the same row space as A. But since B is an echelon matrix, the Lemma tells us that the set of non-$\mathbf{0}$ rows of B is a basis of this space.

Proof of the Theorem. We show how to change A, by means of elementary row operations, into an echelon matrix B. The process yields one row of B at a time; each row results from a three-step process. To get row 1 of B, the steps are as follows.

Step (a)   If $A = 0$, then it is already an echelon matrix (with $r = 0$), so we just take $B = A$.

Step (b)   If $A \neq 0$ let $j(1)$ be the smallest integer $j \in \{1, \ldots, n\}$ such that the jth column of A is not $\mathbf{0}$. If $a_{1,j(1)} \neq 0$, go on to Step (c). If $a_{1,j(1)} = 0$, use an operation of Type I: $R_1 \leftrightarrow R_i$ to bring a non-zero $a_{i,j(1)}$ into the first row of column $j(1)$.

Step (c)   We may assume now that $a_{1,j(1)} \neq 0$. The aim of this step is to make all the other entries in column $j(1)$ equal to zero. If in fact $a_{h,j(1)} \neq 0 (h \neq 1)$, use operation of Type II

$$R_h \to a_{1,j(1)} R_h - a_{h,j(1)} R_1;$$

this gives a new matrix whose $[h, j(1)]$-entry is zero, but whose other entries in column $j(1)$ are as they were before. Repeat this until all entries in this column, except the first, are zero. We now have a matrix† which looks like this:

$$B_1 = \begin{bmatrix} 0 & \dots & 0 & b_{1,j(1)} & * & \dots & * \\ 0 & \dots & 0 & 0 & * & \dots & * \\ 0 & \dots & 0 & 0 & * & \dots & * \end{bmatrix} \Big\} A_1$$

with $b_{1,j(1)} \neq 0$. Row 1 of this matrix will be row 1 of our final echelon matrix B. So we do no more with this row, but apply Steps (a), (b), (c) to the matrix A_1 obtained by removing row 1 from B_1. Step (a): if $A_1 = 0$, then B_1 is already an echelon matrix, and we take $B = B_1$. Step (b): choose smallest $j(2)$ such that column $j(2)$ of A_1 is not zero, etc.

This process goes on until there are no more non-**0** rows left, and then the matrix B we are left with with is echelon. Notice that condition $j(1) < j(2) < \cdots < j(r)$ of **Ech2** is automatically guaranteed by our procedure.

Example 270. The above procedure is easier to do than to describe! The reader is advised to try some of the exercises on this method at the end of this chapter: we give here the working for the matrix

$$A = \begin{bmatrix} 0 & 0 & 1 & -7 & 11 \\ 0 & 2 & 1 & -1 & 3 \\ 0 & 3 & 1 & 2 & -1 \end{bmatrix}.$$

Here $j(1) = 2$. Use $R_1 \leftrightarrow R_2$ to bring 2 into top position in column 2.

$$\begin{bmatrix} 0 & 2 & 1 & -1 & 3 \\ 0 & 0 & 1 & -7 & 11 \\ 0 & 3 & 1 & 2 & -1 \end{bmatrix}$$

---

†Asterisks * indicate entries which have not yet been specified.

Use $R_3 \to 2R_3 - 3R_1$ to make the (3, 2)-entry zero. Note that only row 3 is changed by this operation.

$$B_1 = \begin{bmatrix} 0 & 2 & 1 & -1 & 3 \\ 0 & 0 & 1 & -7 & 11 \\ 0 & 0 & -1 & 7 & -11 \end{bmatrix} \Big\} A_1$$

Since all entries in col 2 are zero (except the first), we have now reached matrix B_1, whose first row will not be altered from now on.

Working on the 2×5 matrix A_1, we find $j(2) = 3$.

Use $R_3 \to R_3 + R_2$ to make (3, 3)-entry zero.

$$B = B_2 = \begin{bmatrix} 0 & 2 & 1 & -1 & 3 \\ 0 & 0 & 1 & -7 & 11 \\ 0 & 0 & 0 & 0 & 0 \end{bmatrix}$$

Since all rows after row 2 are now zero, we are finished.

Conclusion. Row space of A = row space of B has basis $\{(0, 2, 1, -1, 3), (0, 0, 1, -7, 11)\}$. Of course this space has many bases. Different echelon matrices B can easily be obtained from a given matrix A, by using different sequences of row operations. But they must all give the same value for the row rank of A, namely 2.

Example 271. The rows a_1, a_2, a_3 of the matrix A in the last example cannot be linearly independent, since if they were, they would form a basis of row sp A – but row sp A has dimension 2, so this is impossible. In fact $a_1 - 3a_2 + 2a_3 = \mathbf{0}$, which is a non-trivial relation on a_1, a_2, a_3.

9.12 Systems of linear equations

Suppose that $A = (a_{ij})$ is an $m \times n$ matrix over F, and that $h = (h_1, \ldots, h_m)^T$ is a *column m-vector* over F (we denote a column vector as the transpose of a row vector, to save space). Corresponding to A and h is a *system of linear equation.*

$$a_{11}x_1 + \cdots + a_{in}x_n = h_1,$$
$$a_{21}x_1 + \cdots + a_{2n}x_n = h_2, \tag{9.27}$$
$$\vdots$$
$$a_{m1}x_1 + \cdots + a_{mn}x_n = h_m.$$

Any sequence $x_1, \ldots, x_n$ of elements of F such that (9.27) holds is a *solution* of (9.27). If we write this sequence as a column n-vector $x = (x_1, \ldots, x_n)^T$, then (9.27) has a convenient expression in matrix notation, namely

$$Ax = h.$$

For Ax, the product of the $m \times n$ matrix A with the $n \times 1$ matrix x is an $m \times 1$ matrix, i.e. a column m-vector; the ith component of this column m-vector is the term $a_{i1}x_1 + \cdots + a_{in}x_n$ on the left side of the ith equation in (9.27). The system (9.27) is called *homogeneous* if $h_1 = h_2 = \cdots = h_m = 0$; i.e. if h is the zero column m-vector $\mathbf{0}$.

Solution of the system (9.27)

This is done by applying to the equations

$$\mathrm{E}(i): a_{i1}x_1 + \cdots + a_{in}x_n = h_i$$

of system (9.27), the same sequence of elementary operations which we use to change the matrix A to an echelon matrix; the point is, that an elementary operation applied to (9.27) gives an *equivalent* system, i.e. a system with exactly the same solutions $x = (x_1, \ldots, x_n)^T$ as the system (9.27). The elementary operations on equations correspond exactly to the elementary row operations described in the last section.

Type I

Operation $\mathrm{E}(i) \leftrightarrow \mathrm{E}(j)$ simply interchanges the ith equation of the system with the jth. The resulting system has the same solutions as (9.27), clearly.

Type II

Operation $E(i) \to \lambda E(i) + \mu E(j)$ leaves all equations of (9.27) unchanged except the ith, which is replaced by the equation

$$\lambda E(i) + \mu E(j): (\lambda a_{i1} + \mu a_{j1})x_1 + \cdots + (\lambda a_{in} + \mu a_{jn})x_n = \lambda h_i + \mu h_j;$$

here λ, μ are any elements of F such that $\lambda \neq 0$. It is clear that if $x = (x_1, \ldots, x_n)^T$ satisfies (9.27), it satisfies the new system. Conversely, if x satisfies the new system, i.e. if it satisfies $E(h)$ for all $h \neq i$, and also $\lambda E(i) + \mu E(j)$, then it satisfies $E(i) = \lambda^{-1}[\lambda E(i) + \mu E(j)] - \lambda^{-1}E(j)$, and so it satisfies the original system (9.27). Therefore an operation of either Type I or Type II changes (9.27) to an equivalent system.

To solve the system (9.27), the procedure is: apply to (9.27) any sequences of elementary operations, whose corresponding elementary row operations change A into an echelon matrix $B = (b_{ij})$. The resulting system has the form

$$
\begin{aligned}
b_{11}x_1 + \cdots + b_{1n}x_n &= k_1, \\
b_{21}x_1 + \cdots + b_{2n}x_n &= k_2, \\
&\vdots \\
b_{m1}x_1 + \cdots + b_{mn}x_n &= k_m,
\end{aligned}
\tag{9.28}
$$

where $k_1, \ldots, k_m$ are certain linear combinations of $h_1, \ldots, h_m$. Since (9.27) and (9.28) are equivalent systems, we get all solutions $x = (x_1, \ldots, x_n)^T$ of (9.27), by solving (9.28).

We give the result first for a *homogeneous system*.

THEOREM (i) *Suppose $Ax = 0$ is a homogeneous system (9.27) of linear equations, where A is an $m \times n$ matrix of row rank r. Then the solution set $S = \{x \in F^{m,1} | Ax = 0\}$ is a subspace of dimension $n - r$. In particular of $n = r$, then $S = \{0\}$, i.e. the only solution of $Ax = 0$ is $x = 0$.*

Proof. Since all the h_i are zero, the same is true of all the k_i in (9.28), i.e. (9.28) is also a homogeneous system $Bx = 0$. We know that B is an echelon matrix, and so it satisfies conditions **Ech1**, **Ech2** (p. 206); the number r appearing in **Ech1**, **Ech2** is the row rank of B, which is the same as the row rank of A. The last $m - r$ rows of B are zero, and so the last $m - r$ equations (9.28) all take the form $0 = 0$. Since these impose no constraints on $x_1, \ldots, x_n$ we may ignore them; (9.28) is equivalent to the system (9.28′) consisting of the first r equations of (9.28). This system looks as follows

$$b_{1,j(1)}x_{j(1)} + \cdots \qquad\qquad = 0,$$
$$b_{2,j(2)}x_{j(2)} + \cdots \qquad\qquad = 0, \qquad\qquad (9.28')$$
$$\vdots \qquad\qquad\qquad \vdots$$
$$b_{r,j(r)}x_{j(r)} + \cdots = 0,$$

where the matrix entries $b_{i,j(i)}$ $(i = 1, \ldots, r)$ are all non-zero, and $b_{i,j(i)}$ is the first non-zero entry in the ith row of B, starting from the left. The crucial point is: given *arbitrary values* to the $n - r$ components x_j, where $j \in \{1, \ldots, n\}$ but j is not in the set $\{j(1), \ldots, j(r)\}$, there is exactly one solution $x = (x_1, \ldots, x_n)^T$ of the system (9.28'). In fact $x_{j(r)}, x_{j(r-1)}, \ldots, x_{j(1)}$ can be calculated in turn from equations $r, r-1, \ldots, 1$ of (9.28'). To avoid complicated notation, we shall illustrate this in a special case. Let B be the echelon matrix on p. 206. There are $n = 7$ 'variables' x_1, $x_2, \ldots, x_7$, and $j(1) = 3$, $j(2) = 5$, $j(3) = 6$. The system (9.28') consists of $r = 3$ equations

$$x_3 \quad -2x_4 \qquad\quad -4x_6 \quad +7x_7 = 0,$$
$$3x_5 \quad +5x_6 \qquad\quad = 0,$$
$$-8x_6 \quad +2x_7 = 0,$$

where terms with zero coefficient have been omitted. Given arbitrary values x_1, x_2, x_4, x_7 we can use these equations, starting at the bottom, to calculate x_6, x_5, x_3 in turn in terms of x_1, x_2, x_4, x_7:

$$x_6 = \tfrac{1}{4}x_7,$$
$$x_5 = -\tfrac{5}{3}x_6 = -\tfrac{5}{12}x_7,$$
$$x_3 = 2x_4 + 4x_6 - 7x_7 = 2x_4 - 6x_7.$$

So with x_1, x_2, x_4, x_7 given, there is a unique solution x of our system, namely

$$x = (x_1, x_2, 2x_4 - 6x_7, x_4, -\tfrac{5}{12}x_7, \tfrac{1}{4}x_7, x_7)^T. \qquad (9.29)$$

We can write this as a linear combination

$$x = x_1 q_1 + x_2 q_2 + x_4 q_4 + x_7 q_7$$

of the column 7-vectors

$$q_1 = (1, 0, 0, 0, 0, 0, 0)^T,$$

$$q_2 = (0, 1, 0, 0, 0, 0, 0)^T,$$

$$q_4 = (0, 0, 2, 1, 0, 0, 0)^T,$$

and

$$q_7 = (0, 0, -6, 0, -\tfrac{5}{12}, \tfrac{1}{4}, 1)^T.$$

But this means that the solution set S of our system is the subspace of $F^{7,1}$ spanned by the vectors q_1, q_2, q_4, q_7; moreover these vectors form a linearly independent set, because it is clear from (9.29) that $x = x_1 q_1 + x_2 q_2 + x_4 q_4 + x_7 q_7$ has x_1, x_2, x_4, x_7 as its first, second, fourth and seventh components, respectively; hence $x = 0$ only if $x_1 = x_2 = x_4 + x_7 = 0$. Therefore q_1, q_2, q_4, q_7 is a basis of S.

In general, exactly similar calculations will show that the solution set S of the system (9.28$'$) is a subspace of $F^{n,1}$ having a basis $\{q_j\}$, where j runs over the set $\{1, \ldots, n\} \backslash \{j(1), \ldots, j(r)\}$. Since this last set has $n - r$ elements, the dimension of S is $n - r$. Finally (9.28$'$) is equivalent to (9.28), which is equivalent to our original homogeneous system $Ax = 0$. Therefore $Ax = 0$ has the same solution set S. This proves the theorem.

The solution of an *inhomogeneous system*, i.e. a system (9.27) $Ax = h$ with $h \neq 0$, starts off in the same way as for the homogeneous system $Ax = 0$. But we cannot expect the elements $k_1, \ldots, k_m$ appearing in (9.28) to be zero. So the last $m - r$ equations of (9.28) must be taken seriously. They read

$$0 = k_{r+1}, \ldots, 0 = k_m, \tag{9.30}$$

and are called *consistency conditions* for the system $Ax = h$, since *if they do not hold, there is no solution to (9.28), and hence no solution to (9.27) $Ax = h$*. If the consistency conditions do hold, then (9.28) is equivalent to the system (9.28$'$) consisting of the first r equations of (9.28). As before, given arbitrary values x_j for $j \in \{1, \ldots, n\} \backslash \{j(1), \ldots, j(r)\}$, there is exactly one solution of (9.28$'$), which is found by using (9.28$'$) to calculate $x_{j(r)}, \ldots, x_{j(1)}$ in turn.

THEOREM (ii) *Suppose $Ax = h$ is a system (9.29) of linear equations, where A is an $m \times n$ matrix of row rank r. Then there is a set (9.30) of $m - r$ conditions on $h_1, \ldots, h_m$ which must must be satisfied, if (9.27) is to have any solution x at all. If these consistency conditions (9.30) are satisfied, then (9.27) has at least one solution; in particular if $m = r$, then (9.27) has at least one solution.*

Assume that conditions (9.30) are satisfied and that y is any solution of (9.27), $Ay = h$. Then the full solution set S of (9.27) consists of all column m-vectors $x = y + z$, where z is any member of the solution set S_0 of the homogeneous system $Ax = 0$.

Proof. The statements in the first paragraph follow from what we have said above.

Now let y be any solution of (9.27), then an arbitrary element $x \in F^{n,1}$ satisfies (9.27) if and only if $Ax = h \Leftrightarrow Ax = Ay \Leftrightarrow A(x - y) = 0$. Putting $z = x - y$, we have: $x \in S \Leftrightarrow Az = 0 \Leftrightarrow z \in S_0$. So $x \in S$ if and only if $x = y + z$, for some $z \in S_0$.

In practice, it is best to solve the system $Ax = h$ fully by means of the equivalent system (9.28′) (assuming always that the consistency conditions are met); there is no special advantage in solving the homogeneous system $Ax = 0$ first.

Example 272. It is very easy to given an example of an inconsistent (hence insoluble) system: we could take

E(1) $\qquad x_1 + 4x_2 = -1,$

E(2) $\qquad -2x_1 - 8x_2 = 3.$

Using the operation $E(2) \rightarrow E(2) + 2E(1)$ we get the equivalent system

$x_1 + 4x_2 = -1,$

$0x_1 + 0x_2 = 1,$

showing that if the first system had any solution $(x_1, x_2)^T$, it would lead to the contradiction $0 = 1$. This is also very clear from the fact that the left side of E(2) is (-2) times the left side of E(1), hence $E(1) \Rightarrow -2x_1 - 8x_2 = (-2)(-1) = 2$, which contradicts E(2).

Example 273. We go through the solution of the system $Ax = h$, where A is the matrix of Example 270; for the moment, we let $h = (h_1, h_2, h_3)^T$ be an arbitrary column three-vector. Written out, the system $Ax = h$ is

E(1) $\qquad x_3 - 7x_4 + 11x_5 = h_1$

E(2) $\qquad 2x_2 + x_3 - x_4 + 3x_5 = h_2$

E(3) $\qquad 3x_2 + x_3 + 2x_4 - x_5 = h_3$

We now apply to this system, the elementary operations corresponding to the elementary row operations we used in Example 270. It is usual to suppress the variables $x_1, x_2, \ldots, x_5$ during these calculations, so our system of equations again looks like the matrix A, but 'augmented' by a column h at the end

0	0	1	−7	11	h_1
0	2	1	−1	3	h_2
0	3	1	2	−1	h_3

$j(1) = 2$. Use E(1)↔E(2) to bring 2 into top position of column 2. Notice h_1 and h_2 must be interchanged.

0	2	1	−1	3	h_2
0	0	1	−7	11	h_1
0	3	1	2	−1	h_3

Use E(3)→2E(3)−3E(1) to make (3, 2)-entry zero. Note that only equation E(3) is changed by this operation. We have $j(2) = 3$.

0	2	1	−1	3	h_2
0	0	1	−7	11	h_1
0	0	−1	7	−11	$2h_3 - 3h_2$

Use E(3)→E(3)+E(2) to make (3, 3) entry zero.

0	2	1	−1	3	h_2
0	0	1	−7	11	h_1
0	0	0	0	0	$2h_3 - 3h_2 + h_1$

We have brought the matrix B of our system to echelon form. Reintroduce the variables $x_1, \ldots, x_5$ (it happens x_1 does not appear). So the system is

$$2x_2 + x_3 - x_4 + 3x_5 = h_2$$

$$x_3 - 7x_4 + 11x_5 = h_1$$

$$0 = h_1 - 3h_2 + 2h_3.$$

Consistency condition is $h_1 - 3h_2 + 2h_3 = 0$. Unless this is satisfied, the system has no solutions. Assume it is satisfied. Give x_1, x_4, x_5 any values. Then we find from the above that

$$x_3 = h_1 + 7x_4 - 11x_5,$$
$$x_2 = \tfrac{1}{2}(h_2 - x_3 + x_4 - 3x_5)$$
$$= \tfrac{1}{2}(h_2 - h_1 - 6x_4 + 8x_5),$$

hence

$$x = (x_1, \tfrac{1}{2}(h_2 - h_1) - 3x_4 + 4x_5, h_1 + 7x_4 - 11x_5, x_4, x_5)^T,$$

so

(*) $\quad x = y + x_1 q_1 + x_4 q_4 + x_5 q_5,$

where $\quad y = [0, \tfrac{1}{2}(h_2 - h_1), h_1, 0, 0]^T,$

$$q_1 = (1, 0, 0, 0, 0)^T$$
$$q_4 = (0, -3, 7, 1, 0)^T$$
$$q_5 = (0, 4, -11, 0, 1)^T.$$

Thus (*) gives the full solution set, as x_1, x_4, x_5 take all possible values in F. The homogeneous system $Ax = 0$ has the three-dimensional subspace S_0 spanned by $\{q_1, q_4, q_5\}$ as its solution set.

9.13 Matrices and linear maps

The theorems which we have proved on linear equations have interesting interpretations in terms of linear maps. Suppose A is an $m \times n$ matrix over F. We define a map

$$f_A \colon F^{n,1} \to F^{m,1}$$

by the rule: $f_A(x) = Ax$, for all $x \in F^{n,1}$. This map is *linear*: the two conditions **VH1**, **VH2** (see section 9.8) become $A(x + y) = Ax + Ay$, $A(\alpha x) = \alpha(Ax)$, and these hold for all $x, y \in F^{n,1}$ and all $\alpha \in F$, by theorems (i) and (iii) of section 9.10.

Example 274. In case $F = R$, the map f_A has a geometrical interpretation. For this purpose, we regard the points of n-dimensional space as given by *column* n-vectors, i.e. by elements of $R^{n,1}$, rather than by row n-vectors. If $x \neq 0$ is such a point, then $u = f_A(x) = Ax$ is a point in m-dimensional space

$R^{m,1}$. It is possible that $u = 0$, i.e. that $x \in \text{Ker} f_A$, but if $u \neq 0$ then f_A *maps the line* $L(0, x)$ *to the line* $L(0, u)$. For (Example 233, section 9.3) $L(0, x)$ is the set of all λx, $\lambda \in F$, and these points are mapped by f_A to the points $A(\lambda x) = \lambda(Ax) = \lambda u$, for all $\lambda \in F$, i.e. to the points of $L(0, u)$. If x, y are distinct points in $R^{n,1}$, the line $L(x, y)$ joining them is defined to be the set $\{(1-\lambda)x + \lambda y | \lambda \in R\}$ (Example 234, section 9.3). But $f_A[(1-\lambda)x + \lambda y] = (1-\lambda)(Ax) + \lambda(Ay)$, so *if* Ax, Ay *are distinct, then* f_A *maps the line* $L(x, y)$ *to the line* $L(Ax, Ay)$. It is because maps like f_A map lines to lines (in general), that they are called *linear*.

Example 275. Figure 24 illustrates the map $f_A: R^{2,1} \to R^{2,1}$, where

$$A = \begin{bmatrix} 2 & -2 \\ 1 & 2 \end{bmatrix}.$$

This map takes the unit vectors

$$\begin{bmatrix} 1 \\ 0 \end{bmatrix}, \quad \begin{bmatrix} 0 \\ 1 \end{bmatrix}$$

to the points

$$\begin{bmatrix} 2 \\ 1 \end{bmatrix}, \quad \begin{bmatrix} -2 \\ 2 \end{bmatrix}$$

which are the columns of A. The left side of Figure 24 shows some lines in $R^{2,1}$, which are mapped by f_A to the lines on right side.

Kernel and image of f_A

$\text{Ker} f_A$ is, by definition, the set of all $x \in F^{n,1}$ such that $f_A(x) = 0$; therefore $\text{Ker} f_A$ *is the solution set of the homogeneous system of linear equations* $Ax = 0$. We deduce the following theorem.

THEOREM. *If A is an $m \times n$ matrix over F of row rank r, then*

(i) $\text{Ker} f_A$ *is a subspace of $F^{n,1}$ of dimension $n - r$, and*
(ii) f_A *is injective if and only if $n = r$.*

Proof. (i) follows from what we have said above, together with theorem (i), section 9.12.

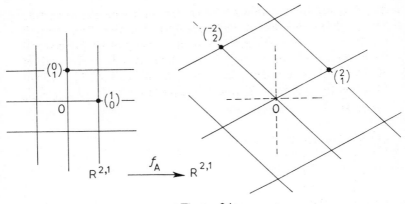

Figure 24

(ii) follows from (i), together with lemma (iv), section 9.8; f_A is injective$\Leftrightarrow$Ker $f_A = \{\mathbf{0}\}\Leftrightarrow$dim Ker $f_A = 0\Leftrightarrow n - r = 0$.

The image of f_A is the column space of A. For Im $f_A = \{Ax \mid x \in F^{n,1}\}$, and if $x = (x_1, \ldots, x_n)^T$ we have

$$Ax = \begin{bmatrix} a_{11}x_1 + \cdots + a_{1n}x_n \\ \vdots \\ a_{m1}x_1 + \cdots + a_{mn}x_n \end{bmatrix} = x_1 \begin{bmatrix} a_{11} \\ \vdots \\ a_{m1} \end{bmatrix}$$

$$+ \cdots + x_n \begin{bmatrix} a_{1n} \\ \vdots \\ a_{mn} \end{bmatrix}.$$

So Im f_A is the set of all linear combinations of the columns of A, i.e. it is the column space of A. We have the

THEOREM. *If A is an $m \times n$ matrix over F of column rank r', then*

(iii) Im f_A *is a subspace of $F^{m,1}$ of dimension r', and*
(iv) f_A *is surjective if and only if $m = r'$.*

Proof. (iii) is immediate from what we have said above. To prove (iv), first notice that dim $F^{m,1} = m$ (the 'unit' column m-vectors $e_1^T, \ldots, e_m^T$ form a basis), and that f_A is surjective if and only if $\mathrm{Im}\, f_A = F_{m,1}$. So if f_A is surjective, then $r' = m$ by (iii). Conversely if $r' = m$, then any basis of $\mathrm{Im}\, f_A$, being a linearly independent subset of $F^{m,1}$ with m elements, must also be a basis of $F_{m,1}$ (Basis Theorem (iii), section 9.7), and this shows that $\mathrm{Im}\, f_A = F_{m,1}$, i.e. f_A is surjective.

Example 276. The 2×2 matrix

$$A = \begin{bmatrix} 2 & -2 \\ 1 & 2 \end{bmatrix}$$

of Example 275 has row rank $r = 2$, and column rank $r' = 2$. It follows from (ii) above that f_A is injective (since $n = r$) and from (iv) that f_A is surjective (since $m = r'$), i.e. f_A is a linear automorphism of $R^{2,1}$. (To prove $r = 2$, note that neither of the rows $x = (2, -2)$, $y = (1, 2)$ is a scalar multiple of the other. This shows that $\{x, y\}$ is linearly independent (Example 247, section 9.6), hence $\{x, y\}$ is basis of row sp A. So $r = \dim$ (row sp A) = 2. A similar argument with the columns of A shows that $r' = 2$.)

Example 277. Let

$$B = \begin{bmatrix} 3 & 2 \\ 6 & 4 \end{bmatrix}.$$

Then $f_B : R^{2,1} \to R^{2,1}$ has as its image the column space of B, and this is the *line* $L(\mathbf{0}, u)$, where

$$u = \begin{bmatrix} 3 \\ 6 \end{bmatrix}$$

(since the second column

$$v = \begin{bmatrix} 2 \\ 4 \end{bmatrix}$$

of B is a scalar multiple of u). This also shows that the column rank r' of B is 1. $\mathrm{Ker}\, f_B$ is the set of those

$$x = \begin{bmatrix} x_1 \\ x_2 \end{bmatrix}$$

such that $Bx = \mathbf{0}$, i.e. such that $3x_1 + 2x_2 = 0$, $6x_1 + 4x_2 = 0$. The second equation follows from the first, so we discard it. Given any value of x_2, say $x_2 = 3\lambda$, the equation $3x_1 + 2x_2 = 0$ gives $x_1 = -2\lambda$; so

$$\text{Ker} f_B = \{(-2\lambda, 3\lambda)^T | \lambda \in R\} = L\left(\mathbf{0}, \begin{bmatrix} -2 \\ 3 \end{bmatrix}\right).$$

This has dimension 1, hence by (i) $n - r = 1$, i.e. $r = n - 1 = 1$. Figure 25 has two pictures of $R^{2,1}$ side by side, one showing $\text{Ker} f_B$ and the other $\text{Im} f_B$.

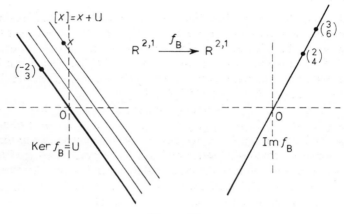

Figure 25

Quotient spaces

Our next ambition is to prove that the row rank r and the column rank r' of any matrix over F are equal. For this we shall need the Homomorphism Theorem for vector spaces, and for *this* we must define quotient spaces.

These are entirely analogous to quotient groups (section 6.8) and quotient rings (section 8.8). Suppose V is a vector space over F and that U is a subspace of V. Since U is then a *subgroup* of the additive group $(V, +)$ (see section 9.4), we may define the additive quotient group V/U, as in section 6.8. Its elements are the cosets $[x] = x + U = \{x + u | u \in U\}$, for all $x \in V$; *addition* is defined by the rule

$$[x] + [y] = [x + y],$$

for all $x, y \in V$. To make V/U into a vector space over F, we define *scalar action* of F on V/U by the rule

$$\alpha[x] = [\alpha x],$$

for all $\alpha \in F$, $x \in V$. Since the elements of V/U are cosets $[x] = x + U$, we must remember the *equality rule* (see theorem, section 6.1)

$$[x] = [x'] \Leftrightarrow x - x' \in U. \tag{9.31}$$

It is easy to verify that these rules of addition and scalar multiplication are consistent with (9.31); it is also easy to verify that with these rules V/U satisfies the vector space axioms **FV1**–**FV8** of section 9.1. The zero element of V/U is $[\mathbf{0}]$, but it is essential to realize that this element may appear in other guises; in fact by (9.31) we have $[x] = [\mathbf{0}]$ if and only if $x \in U$.

Suppose now that V is finitely generated, and that dim $V = n$. From Example 255 (section 9.7) we know that U is also finitely generated, and if dim $U = d$, we have $d \leq n$. We should like to find an expression for dim (V/U). Take any basis $\{x_1, \ldots, x_d\}$ of U. This is a linearly independent subset of V, and so by the Basis Theorem (section 9.7) we can add elements $x_{d+1}, \ldots, x_n$ to $\{x_1, \ldots, x_d\}$ in such a way as to get a basis of V,

$$\{x_1, \ldots, x_d, x_{d+1}, \ldots, x_n\}. \tag{9.32}$$

LEMMA. *The set* $S = \{[x_{d+1}], \ldots, [x_n]\}$ *is a basis for* V/U. *Hence* V/U *is finitely generated, and*

$$dim\ (V/U) = n - d = dim\ V - dim\ U. \tag{9.33}$$

Proof. Take any element $[x] \in V/U$. Since the set (9.32) is a basis of V, we have $x = \alpha_1 x_1 + \cdots + \alpha_n x_n$ for some $\alpha_1, \ldots, \alpha_n \in F$. By the rules of addition and scalar action in V/U, we find $[x] = \alpha_1[x_1] + \cdots + \alpha_n[x_n]$. But if $1 \leq i \leq d$ we have $[x_i] = \mathbf{0}$, because $x_i \in U$. So $[x] = \alpha_{d+1}[x_{d+1}] + \cdots + \alpha_n[x_n]$ is a linear combination of S. This proves $V/U = \mathrm{sp}\ S$.

It remains to show that S is a linearly independent subset of V/U. If $\lambda_{d+1}[x_{d+1}] + \cdots + \lambda_n[x_n] = [\mathbf{0}]$, we have $[v] = [\mathbf{0}]$, where $v = \lambda_{d+1}x_{d+1} + \cdots + \lambda_n x_n$. But $[v] = [\mathbf{0}]$ implies $v \in U$ by the equality rule (9.31), i.e. v is a linear combination of $\{x_1, \ldots, x_d\}$;

say $v = \mu_1 x_1 + \cdots + \mu_d x_d$, with $\mu_1, \ldots, \mu_d \in F$. This equation can be written

$$(-\mu_1)x_1 + \cdots + (-\mu_d)x_d + \lambda_{d+1}x_{d+1} + \cdots + \lambda_n x_n = \mathbf{0}.$$

But since (9.32) is an independent set, all the coefficients in this relation must be zero. In particular $\lambda_{d+1} = \cdots = \lambda_n = 0$; this proves that S is independent, and proves the Lemma.

Example 278. Suppose $V = R^n$, so that U is a subspace of R^n. Let x be any point in R^n. We can get the coset $[x] = x + U$ by applying the translation map T_x (see section 9.3) to U. For example if U is a line through the origin, then the cosets $[x] = x + U$ are all the lines in R^n which are parallel to U. We could take $U = \mathrm{Ker}\, f_B$ on the left side of Figure 25 (here $V = R^{2,1}$), then the quotient space V/U has for its elements all lines parallel to U. By (9.33) above, dim $(V/U) = $ dim $V - $ dim $U = 2 - 1 = 1$. Any coset $[x]$, which is not U itself, forms a basis of V/U.

HOMOMORPHISM THEOREM FOR VECTOR SPACES. *Let V, W be vector spaces over the same field F, and let $f: V \to W$ be a linear map. Then*

  (v) $\mathrm{Ker}\, f$ *is a subspace of V.*
  (vi) $\mathrm{Im}\, f$ *is a subspace of W.*
  (vii) *If we write $U = \mathrm{Ker}\, f$, there is a linear isomorphism*
       $f^*: V/U \to \mathrm{Im}\, f$,

defined by the rule: if $x \in V$, then $f^*([x]) = f(x)$. Here $[x]$ denotes the coset $x + U$. It follows that

$V/\mathrm{Ker}\, f \cong \mathrm{Im}\, f$.

Proof. Parts (v) and (vi) have been proved as lemma (iii), section 9.8. (vii) Since f is a homomorphism between the additive groups $(V, +)$, $(W, +)$ (see Remark 1, section 9.8), we can apply the Homomorphism Theorem for groups (section 7.4) to prove that f^* is well defined, is bijective, and satisfies **VH1** (section 9.8). It remains only to show that f^* satisfies **VH2**. But this is so, since $f^*(\alpha[x]) = f^*([\alpha x]) = f(\alpha x) = \alpha f(x)$ (using **VH2** for f) $= \alpha f^*([x])$, for all $\alpha \in F$, $[x] \in V/U$. The theorem is now proved.

Example 279. In Figure 25, $f = f_B$ maps each point x in the plane $R^{2,1}$ on the left, to point $f(x) = Bx$ in the plane on the right. All points in $U = \text{Ker } f$ map to $\mathbf{0}$, all points in coset $[x] = x + U$ map to the same point $f(x)$ in $\text{Im } f_B$. The isomorphism f^* simply maps $[x]$, regarded now as a single element of V/U, to $f(x)$.

From the Homomorphism Theorem we deduce at once the following theorem.

THEOREM (viii) *Let V, W be finitely generated vector spaces over F, and let $f: V \to W$ be a linear map. Then*

$$\dim \text{Ker } f + \dim \text{Im } f = \dim V.$$

Proof. The Homomorphism Theorem tells us that the quotient space $V/\text{Ker } f$ is isomorphic to $\text{Im } f$, hence these spaces have the same dimension: $\dim (V/\text{Ker } f) = \dim \text{Im } f$. But $\dim (V/\text{Ker } f) = \dim V - \dim \text{Ker } f$ by (9.33), and theorem (viii) follows.

COROLLARY TO THEOREM (viii) *Let A be any $m \times n$ matrix over F, with row rank r and column rank r'. Then $r = r'$.*

Proof. Apply theorem (viii) to the map $f_A : F^{n,1} \to F^{m,1}$, using also theorems (i) and (iii). We get $(n - r) + r' = n$, hence $r = r'$. This proves the Corollary.

Matrix of a linear map

In this section we have studied an $m \times n$ matrix A, by means of the linear map $f_A : F^{m,1} \to F^{m,1}$. It is often useful to go in the opposite direction, and to study linear maps by means of matrices. Suppose V, W are finitely generated vector spaces over F, of dimensions n, m, respectively. Let $f: V \to W$ be a linear map. Take any bases $\{v_1, \ldots, v_n\}$, $\{w_1, \ldots, w_m\}$ of V, W respectively. For each basis element v_j, we write out $f(v_j)$ (which is an element of W) as linear combination of $w_1, \ldots, w_m$. In this way we get equations

$$f(v_j) = a_{1j}w_1 + \cdots + a_{mj}w_m = \sum_{i=1}^{m} a_{ij}w_i, \tag{9.34}$$

for $j = 1, \ldots, n$. The coefficients a_{ij} are elements of F, and so we may form the $m \times n$ matrix $A = (a_{ij})$, which is called *the matrix of f with respect to the bases* $\{v_1, \ldots, v_n\}$, $\{w_1, \ldots, w_m\}$ *of V, W*. Notice that the jth equation (9.34) gives the entries $a_{1j}, \ldots, a_{mj}$ of the jth *column* of A, not of its jth row. We shall denote this matrix $A = M_{v,w}(f)$, to emphasize that it depends not only on the linear map f, but also on the bases $\mathbf{v} = \{v_1, \ldots, v_n\}$ and $\mathbf{w} = \{w_1, \ldots, w_m\}$.

Now suppose we have *any* $m \times n$ matrix A over F. Then *there is a unique linear map* $f: V \rightarrow W$ *such that* $A = M_{v,w}(f)$. In fact we use equations (9.34) to *define* $f(v_j)$ for $j = 1, \ldots, n$; to define $f(x)$ for an arbitrary $x \in V$, we write $x = \xi_1 v_1 + \cdots + \xi_n v_n$ $(\xi_j \in F)$ (x has a unique such expression, because $v_1, \ldots, v_n$ is a basis of V), and we put

$$f(x) = \xi_1 f(v_1) + \cdots + \xi_n f(v_n) = \sum_{j=1}^{n} \sum_{i=1}^{m} a_{ij} \xi_j w_i. \quad (9.35)$$

It is easy to check that the map $f: V \rightarrow W$ defined by (9.35) is linear, and that f also satisfies (9.34), so that the given matrix A is the matrix of f with respect to the bases $\mathbf{v}$, $\mathbf{w}$.

The relation between $f: V \rightarrow W$ and its matrix $A = M_{v,w}(f)$ can be displayed by the following *diagram of linear maps*:

$$
\begin{array}{ccc}
V & \xrightarrow{\quad f \quad} & W \\
\beta_v \downarrow & & \downarrow \beta_w \\
F^{n,1} & \xrightarrow{\quad f_A \quad} & F^{m,1}
\end{array}
\quad (9.36)
$$

The horizontal arrows represent $f: V \rightarrow W$ and $f_A: F^{n,1} \rightarrow F^{m,1}$. The 'vertical' map β_v is the following: if $x \in V$, write $x = \xi_1 v_1 + \cdots + \xi_n v_n$ (as above). Then $\beta_v(x)$ is defined to be the column n-vector $\xi = (\xi_1, \ldots, \xi_n)^T$. Similarly if $y \in W$ and if $y = \eta_1 w_1 + \cdots + \eta_m w_m$ $(\eta_i \in F)$ we define $\beta_w(y) = \eta = (\eta_1, \ldots, \eta_m)^T$. These vertical maps β_v, β_w are linear isomorphisms; in fact β_v is, apart from some differences of notation, the inverse of the isomorphism $f: F^n \rightarrow V$ in the theorem of section 9.8.

The diagram of (9.36) maps is said to *commute* if $f_A \beta_v = \beta_w f$, i.e. if, for any element $x \in V$, the result of applying to x first β_v and then f_A, gives the same result as if we had taken the other 'route' from V to $F^{m,1}$, viz. first f and β_w. The point of diagram (9.36) is that, given $f: V \to W$ and also an $m \times n$ matrix A, then A is the matrix of f with respect to $\mathbf{v}, \mathbf{w}$ if and only if (9.36) commutes. To prove this, note that if $x = \xi_1 v_1 + \cdots + \xi_n v_n$ and if $y = f(x) = \eta_1 w_1 + \cdots + \eta_m w_m$, then, according to our definitions of β_v and β_w, we have

$$f_A \beta_v(x) = f_A(\xi) = A\xi, \qquad and \qquad \beta_w f(x) = \beta_w(y) = \eta. \qquad (9.37)$$

Now A is the matrix of f with respect to $\mathbf{v}, \mathbf{w}$ if and only if (9.35) holds, i.e. if and only if for all x we have

$$\sum_{i=1}^{m} \eta_i w_i = y = f(x) = \sum_{j=1}^{n} \sum_{i=1}^{m} a_{ij} \xi_j w_i.$$

Comparing coefficients of w_i on the two ends of this equation, we get

$$\sum_{j=1}^{n} a_{ij} \xi_j = \eta_i, \qquad i = 1, \ldots, m;$$

but this is equivalent to the single matrix equation $A\xi = \eta$. Going back to (9.37), we get our desired result: $f_A \beta_v = \beta_w f$ if and only if $A = M_{v,w}(f)$.

Since the maps β in (9.36) are both isomorphisms, they have inverses, and we can write the condition $f_A \beta_v = \beta_w f$ as

$$f_A = \beta_w f \beta_v^{-1}, \qquad or \ equivalently \qquad f = \beta_w^{-1} f_A \beta_v. \qquad (9.38)$$

Example 280. If V, W, X are finitely generated vector spaces over F, with bases $\mathbf{v}, \mathbf{w}, \mathbf{x}$ respectively, and if $f: V \to W$, $g: W \to X$ are linear maps whose matrices with respect to these bases are A, B then *the matrix of $gf: V \to X$ is BA.* (It is for this reason that matrix multiplication is defined the way it is!). To prove it, note first that $f_{BA} = f_B f_A$ (since $f_B f_A$ maps any $x \in F^{n,1}$ to $f_B[f_A(x)] = f_B(Ax) = B(Ax) = (BA)x$. Using (9.38) we get $f_{BA} = \beta_x g \beta_w^{-1} \cdot \beta_w f \beta_v^{-1} = \beta_x (gf) \beta_v^{-1}$. But this shows (by (9.38) again) that BA is the matrix of gf with respect to $\mathbf{v}, \mathbf{x}$.

Example 281. Change of basis of a vector space. Suppose we have two bases $\mathbf{v} = \{v_1, \ldots, v_n\}$, $\mathbf{v}' = \{v_1', \ldots, v_n'\}$ of the same vector space V. Let $P = M_{v,v}(\iota_V)$ be the matrix of the identity map $\iota_V: V \to V$ with respect to $\mathbf{v}, \mathbf{v}'$.

By definition, $P = (p_{ij})$ is the $n \times n$ matrix obtained from equations (9.34), when we put $f = \iota_V$, and replace $\mathbf{w}$ by $\mathbf{v}'$, i.e.

$$\iota_V(v_j) = v_j = p_{1j}v_1' + \cdots + p_{nj}v_n', \qquad j = 1, \ldots, n. \tag{9.39}$$

These are the equations which express the basis elements v_j, in terms of the other basis $\mathbf{v}'$. Similarly $Q = M_{\mathbf{v},\mathbf{v}'}(\iota_V)$ is the $n \times n$ matrix $Q = (q_{ij})$ obtained from the equations

$$\iota_V(v_j') = v_j' = q_{1j}v_1 + \cdots + q_{nj}v_n, \qquad j = 1, \ldots, n, \tag{9.40}$$

which express the v_j' in terms of $\mathbf{v}$. Of course if $\mathbf{v} = \mathbf{v}'$, equations (9.39) reduce to $v_j = v_j, j = 1, \ldots, n$. This shows that the matrix $M_{\mathbf{v},\mathbf{v}}(\iota_V) = I_n$, as we might expect. Apply Example 280 in the case $V = W = X, f = g = \iota_V$, $\mathbf{w} = \mathbf{v}'$, $\mathbf{x} = \mathbf{v}$. It gives that $QP = M_{\mathbf{v},\mathbf{v}}(\iota_V) = I_n$. Interchanging the roles of $\mathbf{v}$ and $\mathbf{v}'$, we find $PQ = I_n$. Thus, P, Q are invertible elements of the matrix ring $M_n(F)$, and $Q = P^{-1}$. (We could also deduce this by direct calculation from (9.39) and (9.40).)

Example 282. Let $f: V \to W$ be a linear map, and suppose we have two bases $\mathbf{v}$, $\mathbf{v}'$ of V, and two bases $\mathbf{w}$, $\mathbf{w}'$ of W. We seek a relation between the $m \times n$ matrices $A = M_{\mathbf{v},\mathbf{w}}(f)$ and $B = M_{\mathbf{v}',\mathbf{w}'}(f)$, which are matrices of the *same map* f, but with respect to different pairs of bases. Let $P = M_{\mathbf{v},\mathbf{v}'}(\iota_V)$, and $R = M_{\mathbf{w},\mathbf{w}'}(\iota_W)$. Thus P is the $n \times n$ matrix expressing basis $\mathbf{v}$ in terms of $\mathbf{v}'$ (see (9.39), above); similarly R is the $m \times m$ matrix expressing basis $\mathbf{w}$ in terms of $\mathbf{w}'$. By the last example, P and R are both invertible matrices.

Now apply Example 280 twice, as follows. First take V, W, X to be V, W, W, use bases $\mathbf{v}$, $\mathbf{w}$, $\mathbf{w}'$ of these respective spaces, and take $g = \iota_W$. We get $M_{\mathbf{v},\mathbf{w}'}(f) = RA$. Next take V, W, X to be V, V, W, use bases $\mathbf{v}$, $\mathbf{v}'$, $\mathbf{w}'$ and take $f = \iota_V$, $g = f$. We get $M_{\mathbf{v},\mathbf{w}'}(f) = BP$. Hence $RA = BP$, from which we get the formula $B = RAP^{-1}$.

A very common special case is where $V = W$. Taking $\mathbf{v} = \mathbf{w}$, $\mathbf{v}' = \mathbf{w}'$ we have: If $f: V \to V$ is a linear map, and if $A = M_{\mathbf{v},\mathbf{v}}(f)$ and $B = M_{\mathbf{v}',\mathbf{v}'}(f)$ are the matrices of f with respect to $\mathbf{v}$, $\mathbf{v}$ and $\mathbf{v}'$, $\mathbf{v}'$, where $\mathbf{v}$ and $\mathbf{v}'$ are bases of V, then $B = PAP^{-1}$ (P being the matrix which expresses basis $\mathbf{v}$ in terms of $\mathbf{v}'$).

9.14   Invertible matrices. The group $GL_n(F)$

Suppose that A is a square matrix, say an $n \times n$ matrix over F. Then A is an element of the ring $M_n(F)$ (section 9.10). In accordance with the general definitions of ring theory, A is said to be *invertible* if there is some $B \in M_n(F)$ such that $AB = BA = I_n$; in this case B is the *inverse* of A, and we write $B = A^{-1}$. The set $U[M_n(F)]$ of all invertible $A \in M_n(F)$ is a multiplicative group (see theorem, section 8.1) with the identity matrix I_n as its identity element. This group is called the *general linear group of degree n over F*, and is denoted $GL_n(F)$.

If we want to know whether a given $n \times n$ matrix A is invertible, we can ask if it is possible to find an $n \times n$ matrix B such that $AB = I_n$. This is a *matrix equation*, in the sense that the 'unknown' term B is a matrix. But we can reduce it to n systems of linear equations, by looking at the columns of both sides. For this purpose the following elementary lemma is useful.

LEMMA. *Let A, B be matrices of shapes (m, n), (n, p) respectively. Let $b^{(1)}, \ldots, b^{(p)}$ be the columns of B. Then the columns of AB are the column m-vectors $Ab^{(1)}, \ldots, Ab^{(p)}$.*

Proof. This is little more than a re-statment of the rule for matrix multiplication. Let $c^{(1)}, \ldots, c^{(p)}$ be the columns of the $m \times p$ matrix $AB = C = (c_{ik})$. For given $i \in \{1, \ldots, m\}$, $k \in \{1, \ldots, p\}$, the ith component of $Ab^{(k)}$ is (ith row of A) times $b^{(k)}$. But (see (9.19), section 9.9) this is exactly the same as c_{ik}, which is the ith component of the column vector $c^{(k)}$. This shows that $Ab^{(k)} = c^{(k)}$ for $k = 1, \ldots, p$, and thus proves the lemma.

Going back to the $n \times n$ matrices A, B and I_n, notice that the columns I_n are the 'unit column n-vectors' $e_1^T = (1, 0, \ldots, 0)^T$, $e_2^T = (0, 1, \ldots, 0)^T, \ldots, e_n^T = (0, 0, \ldots, 1)^T$. So by the Lemma, $AB = I_n$ if and only if

$$Ab^{(1)} = e_1^T, \qquad Ab^{(2)} = e_2^T, \ldots, Ab^{(n)} = e_n^T. \tag{9.41}$$

We shall now give two criteria for A to be invertible.

THEOREM. *Let A be an $n \times n$ matrix over F. Then*
  (i) *A is invertible if and only if rank $A = n$.*
  (ii) *If B is an $n \times n$ matrix such that $AB = I_n$, then also $BA = I_n$, hence A is invertible and $B = A^{-1}$.*

Proof. Parts (i) and (ii) will follow from two further statements (a) and (b).

(a) *If rank $A = n$ then there exists $B \in M_n(F)$ such that $AB = I_n$.* For if rank $A = n$, then for any given $h \in F^{n,1}$ there is some $x \in F^{n,1}$ with $Ax = h$ (theorem (ii), section 9.12; we have $n = m = r$). So we can find column vectors $b^{(1)}, \ldots, b^{(n)}$ to satisfy (9.41). Let B be the $n \times n$ matrix whose columns are $b^{(1)}, \ldots, b^{(n)}$. By what we said above, $AB = I_n$.

(b) *If $B \in M_n(F)$ is any matrix satisfying $AB = I_n$, then rank $B = n$.* For the columns $b^{(1)}, \ldots, b^{(n)}$ satisfy (9.41). But this shows that they form a linearly independent set. For if $\lambda_1 b^{(1)} + \cdots + \lambda_n b^{(n)} = \mathbf{0}$, we multiply this equation by A on the left, and get $\lambda_1 e_1^T + \cdots + \lambda_n e_n^T = \mathbf{0}$, i.e. $(\lambda_1, \ldots, \lambda_n)^T = \mathbf{0}$. So $\lambda_1 = \cdots = \lambda_n = 0$. Any linearly independent subset of $F^{n,1}$, which contains n elements, is a basis of $F^{n,1}$ (Basis Theorem (iii), section 9.7). Hence col sp $B = F^{n,1}$, showing that the column rank of B is n, hence (p. 223) the rank of B is n.

Proof of (ii). If $AB = I_n$, then by (b) rank $B = n$, so by (a) (applied to B in place of A) there is some $C \in M_n(F)$ such that $BC = I_n$. Then $A = AI_n = A(BC) = (AB)C = I_nC = C$. So $BA = I_n$, showing that A is invertible and $B = A^{-1}$.

Proof of (i). If rank $A = n$, then by (a) there is some B with $AB = I_n$, and by (b) there must also hold $BA = I_n$; thus A is invertible. Now suppose A is invertible. There must be some B with $AB = BA = I_n$. Apply (b) *with the roles of A and B interchanged*. We get that rank $A = n$. This proves the theorem.

Calculation of inverse matrix

We shall give two methods, both based on the technique of solving systems of linear equations (see Example 273, section 9.12), to calculate the inverse A^{-1} of a given $n \times n$ matrix A of rank n. (Since this technique works by reducing A to an equivalent echelon matrix, we automatically check that rank A is actually n.)

Method 1. Use the method of Example 273 to solve the system $Ax = y$, with $y = (y_1, \ldots, y_n)^T$ an arbitrary column n-vector; then write out the solution $x = (x_1, \ldots, x_n)^T$ as a system of equations

$x_i = b_{i1}y_1 + \cdots + b_{in}y_n$ $(i = 1, \ldots, n)$. Since $Ax = y \Rightarrow A^{-1}Ax = A^{-1}y \Rightarrow I_n x = A^{-1}y \Rightarrow x = A^{-1}y$, the matrix (b_{ij}) which appears in these equations, is A^{-1}.

Method 2. Use the method of Example 273 to solve the n systems of equations (9.41): the solution $b^{(j)}$ of the system $Ab^{(j)} = e_j^T$ is the jth column of the desired matrix $B = A^{-1}$. It is usual to solve the n systems in one process, by putting the vectors $e_1^T, \ldots, e_n^T$ together into a single matrix I_n, which is written to the right of A, giving an $n \times (2n)$ matrix (A, I_n). Apply row operations to this matrix, so as to change A into an echelon matrix, and then to I_n. By this time the right hand matrix will have been changed to A^{-1}. The methods are illustrated in the next two examples.

Example 283. Find the inverse of

$$A = \begin{bmatrix} 0 & 1 & -2 \\ 3 & 1 & 0 \\ 2 & 0 & -1 \end{bmatrix},$$

by method 1, i.e. by solving the system of equations $Ax = y$:

$$\begin{aligned} x_2 - 2x_3 &= y_1 \\ 3x_1 + x_2 \phantom{{}-2x_3} &= y_2, \\ 2x_2 \phantom{{}+x_2} - x_3 &= y_3. \end{aligned}$$

We use the method of Example 273 (section 9.12), which is to change the system by elementary operations on equations, to a system $Bx = z$, where B is an echelon matrix. Suppressing the variables x_1, x_2, x_3 for the moment, we work on the 'augmented' matrix

0	1	-2	y_1
3	1	0	y_2
2	0	-1	y_3

To get non-zero entry at top of column 1, use E(1)⇔E(2).

3	1	0	y_2
0	1	-2	y_1
2	0	-1	y_3

Make (3, 1)-entry zero, by using E(3)→3E(3)−2E(1).

$$\begin{array}{cccc} 3 & 1 & 0 & y_2 \\ 0 & 1 & -2 & y_1 \\ 0 & -2 & -3 & 3y_3 - 2y_2 \end{array}$$

Make (3, 2)-entry zero, by using $E(3) \to E(3) + 2E(2)$.

$$\begin{array}{cccc} 3 & 1 & 0 & y_2 \\ 0 & 1 & -2 & y_1 \\ 0 & 0 & -7 & 3y_3 - 2y_2 + 2y_1 \end{array}$$

Now the 3×3 matrix B on the left is in echelon form. Bring back the variables x_1, x_2, x_3 and solve the resulting system.

$$\begin{aligned} 3x_1 + x_2 \quad\;\; &= y_2, \\ x_2 - 2x_3 &= y_1, \\ -7x_3 &= 2y_1 - 2y_2 + 3y_3. \end{aligned}$$

This gives

$$x_3 = -\tfrac{2}{7}y_1 + \tfrac{2}{7}y_2 - \tfrac{3}{7}y_3,$$
$$x_2 = y_1 + 2x_3 = \tfrac{3}{7}y_1 + \tfrac{4}{7}y_2 - \tfrac{6}{7}y_3$$
$$x_1 = \tfrac{1}{3}(y_2 - x_2) = \tfrac{1}{3}(-\tfrac{3}{7}y_1 + \tfrac{3}{7}y_2 + \tfrac{6}{7}y_3),$$

and re-writing these in natural order

$$x_1 = -\tfrac{1}{7}y_1 + \tfrac{1}{7}y_2 + \tfrac{2}{7}y_3,$$
$$x_2 = \tfrac{3}{7}y_1 + \tfrac{4}{7}y_2 - \tfrac{6}{7}y_3,$$
$$x_3 = \tfrac{2}{7}y_1 + \tfrac{2}{7}y_2 - \tfrac{3}{7}y_3,$$

from which we read off

$$A^{-1} = \begin{bmatrix} -\tfrac{1}{7} & \tfrac{1}{7} & \tfrac{2}{7} \\[4pt] \tfrac{3}{7} & \tfrac{4}{7} & -\tfrac{6}{7} \\[4pt] -\tfrac{2}{7} & \tfrac{2}{7} & -\tfrac{3}{7} \end{bmatrix}$$

Example 284. Same matrix A as above. Find inverse by method 2, i.e. use now operations on the matrix below, until the left-hand 3×3 matrix becomes I_3.

$$\begin{array}{cccccc} 0 & 1 & -2 & 1 & 0 & 0 \\ 3 & 1 & 0 & 0 & 1 & 0 \\ 2 & 0 & -1 & 0 & 0 & 1 \end{array}$$

$R_1 \leftrightarrow R_2$.

$$\begin{array}{cccccc} 3 & 1 & 0 & 0 & 1 & 0 \\ 0 & 1 & -2 & 1 & 0 & 0 \\ 2 & 0 & -1 & 0 & 0 & 1 \end{array}$$

$R_3 \rightarrow 3R_3 - 2R_1$.

$$\begin{array}{cccccc} 3 & 1 & 0 & 0 & 1 & 0 \\ 0 & 1 & -2 & 1 & 0 & 0 \\ 0 & -2 & -3 & 0 & -2 & 3 \end{array}$$

$R_3 \rightarrow R_3 + 2R_2$.

$$\begin{array}{cccccc} 3 & 1 & 0 & 0 & 1 & 0 \\ 0 & 1 & -2 & 1 & 0 & 0 \\ 0 & 0 & -7 & 2 & -2 & 3 \end{array}$$

So far, we have used the same operations as in the previous Example. Now use $R_1 \rightarrow R_1 - R_2$ to make (1, 2)-entry zero.

$$\begin{array}{cccccc} 3 & 0 & 2 & -1 & 1 & 0 \\ 0 & 1 & -2 & 1 & 0 & 0 \\ 0 & 0 & -7 & 2 & -2 & 3 \end{array}$$

Use $R_1 \rightarrow 7R_1 + 2R_3$ to make (1, 3)-entry zero, then $R_2 \rightarrow 7R_2 - 2R_3$ to make (2, 3)-entry zero.

$$\begin{array}{cccccc} 21 & 0 & 0 & -3 & 3 & 6 \\ 0 & 7 & 0 & 3 & 4 & -6 \\ 0 & 0 & -7 & 2 & -2 & 3 \end{array}$$

Finally $R_1 \rightarrow \frac{1}{21}R_1$, $R_2 \rightarrow \frac{1}{7}R_2$, $R_3 \rightarrow -\frac{1}{7}R_3$.

$$\begin{array}{cccccc} 1 & 0 & 0 & -\frac{1}{7} & \frac{1}{7} & \frac{2}{7} \\ 0 & 1 & 0 & \frac{3}{7} & \frac{4}{7} & -\frac{6}{7} \\ 0 & 0 & 1 & -\frac{2}{7} & \frac{2}{7} & -\frac{3}{7} \end{array}$$

The inverse matrix A^{-1} appears on the right.

It should be emphasized that there are many ways of carrying out these calculations. For example one could reduce the left-hand 3×3 matrix in Example 283 to the identity matrix, as we have done above. There is no difference in principle between Methods 1 and 2.

Subgroups of $GL_n(F)$

If F is an infinite field (as, for example, $F = R$ or $F = C$) then the group $GL_n(F)$ of all invertible $n \times n$ matrices A over F is infinite. However, it has many interesting subgroups, some of which are finite. We give some examples next. In these examples we take $F = R$.

Example 285. Rotations in the plane. Each invertible matrix $A \in GL_2(R)$ gives rise to a linear map $f_A: R^{2,1} \to R^{2,1}$ (see Example 274, section 9.13). Because rank $A = 2$, the map f_A is bijective (this follows from theorems (ii), (iv) of section 9.13. Or it can be proved directly that f_A has inverse f_B, where B is the inverse of A). The map f_A is completely determined by

$$f_A \begin{bmatrix} 1 \\ 0 \end{bmatrix} \quad \text{and} \quad f_A \begin{bmatrix} 0 \\ 1 \end{bmatrix};$$

in fact these are the column vectors of A. If we interpret $R^{2,1}$ as the plane, this means that f_A is determined by these two points

$$f_A \begin{bmatrix} 1 \\ 0 \end{bmatrix}, \quad f_A \begin{bmatrix} 0 \\ 1 \end{bmatrix},$$

together with the fact that f_A maps each line or $R^{2,1}$ to a line (Example 274).

Figure 26 illustrates the case

$$A = \begin{bmatrix} \cos \theta & -\sin \theta \\ \sin \theta & \cos \theta \end{bmatrix},$$

where θ is some fixed real number. f_A takes

$$\begin{bmatrix} 1 \\ 0 \end{bmatrix} \quad \text{to} \quad \begin{bmatrix} \cos \theta \\ \sin \theta \end{bmatrix} \quad \text{and} \quad \begin{bmatrix} 0 \\ 1 \end{bmatrix} \quad \text{to} \quad \begin{bmatrix} -\sin \theta \\ \cos \theta \end{bmatrix},$$

and it is called the (*anti-clockwise*) *rotation of the plane through an angle* θ, with **0** as fixed point. In Figure 26, the column two-vectors are written as row vectors to save space; the solid lines are the images under f_A of the broken lines.

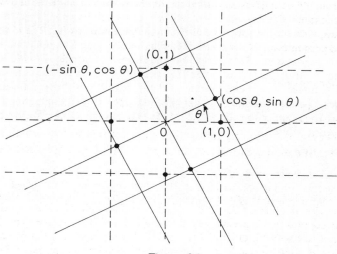

Figure 26

Example 286. Reflection. The linear map $f_J: R^{2,1} \to R^{2,1}$, where

$$J = \begin{bmatrix} 1 & 0 \\ 0 & -1 \end{bmatrix},$$

takes

$$x = \begin{bmatrix} x_1 \\ x_2 \end{bmatrix} \quad \text{to} \quad Jx = \begin{bmatrix} x_1 \\ -x_2 \end{bmatrix},$$

so that x is mapped to its reflection in the line

$$L\left(\mathbf{0}, \begin{bmatrix} 1 \\ 0 \end{bmatrix}\right),$$

i.e. in the x_1-axis. This map is called the *reflection* in this line. Notice that $J^2 = I_2$.

Example 287. Rotation and orthogonal groups. Write the matrix A of Example 285 as $A(\theta)$. We can check that $A(\theta)A(\phi) = A(\theta + \phi)$, $A(0) = I_2$ and $A(\theta)^{-1} = A(-\theta)$, for any θ, $\phi \in R$. (The last equation is most easily proved by showing that $A(\theta)A(-\theta) = I_2$. It follows by theorem (ii) of this section that $A(-\theta)$ is the inverse of $A(\theta)$.) Thus the set $\{A(\theta) | \theta \in R\}$ is a subgroup of $GL_2(R)$; this is called the *rotation group of degree 2*, and we shall denote it $0_2^+(R)$.

Now consider the subgroup S of $GL_2(R)$ generated by $0_2^+(R)$ and the reflection matrix J of Example 286. This means that S consists of all 'words' (see section 5.3) which can be made from the matrices $A(\theta)$ ($\theta \in R$), together with J; i.e. we must take all matrices $u_1^{m_1} u_2^{m_2} \ldots u_f^{m_f}$, where f is any positive integer, and each u_i is either $A(\theta)$ for some θ, or is J; the m_i are any integers. But because $A(\theta)^{m_i} = A(m_i\theta)$, while, since $J^2 = I_2$, any power J^{m_i} of J is equal either to J or to I_2, we see that our 'word' can always be reduced to the form

$$A(\theta_1)JA(\theta_2) \ldots JA(\theta_n),$$

$\theta_1, \ldots, \theta_n$ being elements of R, n some positive integer. Now we can reduce this still further, because $A(\theta)J = JA(-\theta)$ for any $\theta \in R$ (check by direct calculation). Therefore the expression above can be written in one of the forms

$$A(\theta) = \begin{bmatrix} \cos\theta & -\sin\theta \\ \sin\theta & \cos\theta \end{bmatrix}$$

or

$$A(\theta)J = \begin{bmatrix} \cos\theta & \sin\theta \\ \sin\theta & -\cos\theta \end{bmatrix}.$$

So our subgroup $S = 0_2^+ \cup 0_2^+ . J$ (writing 0_2^+ for $0_2^+(R)$), i.e. S is the union of two *cosets* 0_2^+, $0_2^+ . J$ of 0_2^+. Moreover these cosets are disjoint, since otherwise we should have $0_2^+ = 0_2^+ . J$, which would imply that $J \in 0_2^+$; but it is easy to check that there is no $\theta \in R$ such that $J = A(\theta)$. So S is the union of two disjoint sets $0_2^+(R)$ and $0_2^+(R) . J$. S is called the *orthogonal group of degree 2*, and is denoted $0_2(R)$. For definition of the orthogonal group $0_n(R)$ for general $n \geq 1$, see exercises at the end of this chapter.

Example 288. Dihedral group. Let n be a positive integer, and let $A = A(2\pi/n)$. A produces a rotation of the plane through angle $2\pi/n$. Hence $A^n = I_2 = I$, while $I, A, A^2, \ldots, A^{n-1}$ are distinct matrices. These elements form a cyclic subgroup of order n of $GL_2(R)$, which we denote $C(n)$. The set $D(2n) = C(n) \cup C(n)J$, consisting of the disjoint subsets $C(n)$, $C(n)J$, is also a subgroup of $GL_2(R)$. In fact we find by direct calculation that $JA^j = A^{-j}J$ ($j \in Z$), for which it follows that all of the products $A^i . A^j$, $A^i . A^jJ$, $A^iJ . A^j$, $A^iJ . A^jJ$ lie in $D(2n)$ (e.g. $A^iJ . A^jJ = A^iA^{-j}JJ = A^{i-j}$), and that the inverse of A^iJ is A^iJ itself.

This group $D(2n)$ is called the *dihedral group of order 2n*. The multiplication table of $D(8)$ is given at the end of the book.

Example 289. Quaternion group of order 8. Let I, P, Q, R be the matrices of Example 268 (p. 203). Then $Q(8) = \{I, -I, P, -P, Q, -Q, R, -R\}$ is a subgroup of $GL_2(C)$ of order 8 – this follows from the rules for multiplying P, Q, R (see Example 268). The multiplication table of $Q(8)$ is given at the end of the book. Note $D(8) \not\cong Q(8)$, because $D(8)$ has five elements of order 2, while $Q(8)$ has only one.

Determinants

There is a useful method for deciding if a square matrix A is invertible, based on the idea of the *determinant of A*. This is an element of F, written det A or $|A|$, and defined as follows

$n = 1$

Here $A = (a_{11})$, and det A is simply a_{11}.

$$n = 2 \quad \det \begin{bmatrix} a_{11} & a_{12} \\ a_{21} & a_{22} \end{bmatrix} = a_{11}a_{22} - a_{12}a_{21}.$$

$$n = 3 \quad \det \begin{bmatrix} a_{11} & a_{12} & a_{13} \\ a_{21} & a_{22} & a_{23} \\ a_{31} & a_{32} & a_{33} \end{bmatrix} =$$

$a_{11}a_{22}a_{33} + a_{13}a_{21}a_{32} + a_{12}a_{23}a_{31} - a_{11}a_{23}a_{32}$

$- a_{13}a_{22}a_{31} - a_{12}a_{21}a_{33}.$

In general, det A is a sum of $n!$ terms of form

$$\varepsilon(\pi)a(\pi) = \pm a_{1,\pi(1)}a_{2,\pi(2)} \cdots a_{n,\pi(n)}, \tag{9.42}$$

where π runs over all the elements of the symmetric group $S(n)$, and $\varepsilon(\pi) = +1$ or -1, according as π is even or odd (Example 152, p. 93).

Example 290. det $I_n = 1$ (all terms (9.42) vanish except the term $a_{11}a_{22} \cdots a_{nn}$).

Example 291. Let

$$A = \begin{bmatrix} \cos\theta & -\sin\theta \\ \sin\theta & \cos\theta \end{bmatrix}$$

(Example 285). Then det $A = \cos^2\theta - (-\sin^2\theta) = \cos^2\theta + \sin^2\theta = 1$. If we take the matrix

$$AJ = \begin{bmatrix} \cos\theta & \sin\theta \\ \sin\theta & -\cos\theta \end{bmatrix},$$

we find det $(AJ) = -1$. This gives a proof that the cosets $0_2^+(R)$, $0_2^+(R)J$ of Example 287 are disjoint, since all elements of $0_2^+(R)$ have determinant $+1$, and all elements of $0_2^+(R)J$ have determinant -1.

Two fundamental facts about determinants are the following. Let A, B be $n \times n$ matrices over F. Then

(iii) *A has rank n (i.e. is invertible) if and only if det $A \neq 0$.*
(iv) det $(AB) = $ det A . det B.

Space does not permit us to prove these facts here. For a good account of determinants, see P. M. Cohn, *Linear Equations*, Routledge & Kegan Paul, London. But we shall outline the proofs of (iii) and (iv) in case $n = 2$.

Fact (iv) can be proved by direct calculation. By Example 290 one sees that det $I_2 = 1$. Suppose $A \in M_2(F)$ has inverse B, then from (iv) and the fact that $AB = I_2$, one has det A . det $B = 1$. But this forces det A to be non-zero. Conversely, suppose det $A \neq 0$; we want to prove that A is invertible. Let

$$A = \begin{bmatrix} a_1 & b_1 \\ a_2 & b_2 \end{bmatrix},$$

and define

$$B = \frac{1}{\Delta} \begin{bmatrix} b_2 & -b_1 \\ -a_2 & a_1 \end{bmatrix},$$

where $\Delta = $ det $A = a_1 b_2 - a_2 b_1$.

Direct calculation gives $AB = I_2 = BA$. So we have proved A is invertible if and only if det $A \neq 0$. Taken together with theorem (i) in this section, this proves (iii).

Exercises for Chapter 9

1 If x, y are elements of a vector space V over the field F, and if $\alpha \in F$, $\alpha \neq 0$, show that $\alpha x = y \Leftrightarrow x = \alpha^{-1} y$.

2 Prove theorem (iii) of section 9.1: $(-\alpha)v = -\alpha v = \alpha(-v)$, for all $\alpha \in F$, $v \in V$.

3 If F is a field, show that F can be regarded as a vector space over F, using the operations of F (as field) to give addition and scalar action for $V = F$ (as vector space).

4 Let $a = (-1, 4)$, $b = (2, 1)$, $c = (1, 3)$, $d = (-2, 2)$ be elements of R^2, interpreted as points in the plane. Describe the points in the lines $L(a, b)$ and $L(c, d)$. Find the intersection of these lines.

5 Let l, m, n be the midpoints of the sides ab, $0b$, $0a$ of a triangle $0ab$ (regard these points as elements of R^2). Prove that the lines $0l$, am, bn have a common point.

6 Let $\alpha = (\alpha_1, \alpha_2, \alpha_3)$ be a non-0 element of R^3, and let Π_α be the set of all points $x \in R^3$ which satisfy $\alpha_1 x_1 + \alpha_2 x_2 + \alpha_3 x_3 = 0$. Prove that if u, v are any two distinct points of Π_α, then every point of the line $L(u, v)$ lies on Π_α.

7 Let a, b be real numbers, not both zero, and let Λ be the set of all points $(x, y) \in R^2$ which satisfy $ax + by = 0$. Prove that Λ is a line, as defined in Example 233, p. 172.

8 If Λ is the set of the last exercise, and if $\Lambda' = \{(x, y) \in R^2 \mid ax + by = h\}$ (h being a fixed real number), find a translation map T_u on R^2 which maps Λ onto Λ'.

9 Show that the set T of all translations $T_x (x \in R^n)$ is a subgroup of the symmetric group $S(R^n)$.

10 Let A be any set and E any vector space over F. Prove that the set $M = M(A, E)$ of all maps $f : A \to E$ becomes a vector space over F, if the operations of addition and scalar action are defined as follows: If f, $g \in M$ define $f + g \in M$ by $(f + g)(a) = f(a) + g(a)$, all $a \in A$. If $\alpha \in F$ and $f \in M$, define $\alpha f \in M$ by $(\alpha f)(a) = \alpha f(a)$. [See Example 230, p. 169.]

11 Show that the intersection $\cap \mathcal{U}$ of a set $\mathcal{U}$ of subspaces of a given vector space V, is a subspace of V.

12 If U, U' are subspaces of V, show that $U + U'$ is the intersection of the set $\mathcal{U}$ of all subspaces S of V, which contain both U and U'.

13 Prove that the vectors $x = (0, 1, -1)$, $y = (0, 1, 1)$ of R^3 span the subspace $\Pi = \{(u_1, u_2, u_3) | u_1 = 0\}$ of R^3. Prove that $\{x, y\}$ is linearly independent. Find the dimension of Π.

14 Find vectors x, y in the subspace $\Pi = \{(u_1, u_2, u_3) | u_1 - u_2 + u_3 = 0\}$ of R^3, such that $\{x, y\}$ is linearly independent. Prove that $\text{sp}\{x, y\} = \Pi$. Find dim Π.

15 Prove that the elements $x = (x_1, x_2)$, $y = (y_1, y_2)$ of F^2 form a linearly dependent set, if and only if $x_1 y_2 = x_2 y_1$. [Use Example 247, p. 181.]

16 Generalize the last exercise as follows: prove that $x = (x_1, \ldots, x_n)$, $y = (y_1, \ldots, y_n)$ form a linearly dependent subset of F^n, if and only if $x_i y_j = x_j y_i$, for all $i, j \in \{1, 2, \ldots, n\}$ such that $i < j$.

17 *The vector space $\text{Hom}_F(V, W)$.* If V, W are vector spaces over F, let $\text{Hom}_F(V, W)$ denote the set of all linear maps $f: V \to W$. Prove that $\text{Hom}_F(V, W)$ is a subspace of the vector space $M(V, W)$ of *all* maps $f: V \to W$ (see Exercise 10).

18 Prove that the map $f: F^m \to F^n$ (m, n are positive integers with $m \geq n$) given by the rule: $f[(x_1, \ldots, x_m)] = (x_1, \ldots, x_n)$, is linear. Find Ker f and Im f.

19 *Symmetric and antisymmetric matrices.* Let $A = (a_{ij})$ be an $n \times n$ matrix over F. *Definitions:* A is *symmetric* if $A = A^T$ and *antisymmetric* if $A^T = -A$.

Prove that if U is any $m \times n$ matrix, then $A = U^T U$ is symmetric.

20 (Notation of preceding exercise.) Prove that the sets $\mathscr{S}$ and $\mathscr{A}$ of symmetric and antisymmetric matrices are both subspaces of the vector space $F^{n \times n} = M_n(F)$. If $F = R$, find bases of $\mathscr{S}$ and $\mathscr{A}$, respectively, and prove that dim $\mathscr{S} = \frac{1}{2} n(n + 1)$, dim $\mathscr{A} = \frac{1}{2} n(n - 1)$.

21 Let $\{E_{ij} | i, j = 1, \ldots, n\}$ be the basis of $M_n(F) = F^{n,n}$ given in Example 259 (p. 194). Prove $E_{ij} E_{kl} = \delta_{jk} E_{il}$, for all $i, j, k, l \in \{1, \ldots, n\}$, where the symbol δ_{jk} (often called the *Kronecker delta*) stands for 0 if $j \neq k$, and for 1 if $j = k$. (Try this first for $n = 2$.)

22 *Ideals of $M_n(F)$.* Prove that $M_n(F)$ is a simple ring (see section 8.7). [We have $I_n \neq 0_{n,n}$, so all we need is to prove that any ideal J of $M_n(F)$ which is not $\{0\}$, is the whole of $M_n(F)$. Take any non-zero matrix $A = (a_{ij})$ in J. Since $A \neq 0$, there is

some pair of suffixes k, l such that $a_{kl} \neq 0$. Use the fact that J is an ideal, together with Exercise 21, to prove that J contains $a_{kl}^{-1} E_{ik} A \, E_{lj} = E_{ij}$, for all $i, j \in \{1, \ldots, n\}$.]

23 If
$$A = \begin{pmatrix} a_1 & b_1 \\ a_2 & b_2 \end{pmatrix} \in M_2(F),$$
prove that $A^2 - tA + dI_2 = 0$, where $t = a_1 + b_2$ and $d = a_1 b_2 - a_2 b_1$.

24 Let $A \in F^{m,n}$, $B \in F^{n,p}$. Prove that each row of the $m \times p$ matrix $C = AB$, is a linear combination of the rows of B. Hence prove row sp $AB \subseteq$ row sp B, and that row rank of $AB \leq$ row rank of B.

25 Find the dimension of the subspace S of R^4 spanned by the vectors $(1, -1, 1, -1)$, $(3, 0, 2, 4)$, $(0, -1, 1, -5)$, $(1, 1, 1, 1)$. [Take these as rows of a 4×4 matrix A; use method of Example 270, p. 208.]

26 Find necessary and sufficient conditions on h_1, h_2, h_3, h_4, so that the system of equations below has a solution
$$\begin{aligned}
x_1 - x_2 + x_3 - x_4 &= h_1 \\
3x_1 \phantom{{}-x_2} + 2x_3 + 4x_4 &= h_2 \\
-x_2 + x_3 - 5x_4 &= h_3 \\
x_1 + x_2 + x_3 + x_4 &= h_4.
\end{aligned}$$

Find a basis for the solution space of the homogeneous system.

27 Find the row rank of
$$A = \begin{bmatrix} 1 & -1 \\ -2 & 2 \\ 3 & 4 \end{bmatrix}.$$

Show that the system $Ax = h$ [$h = (h_1, h_2, h_3)^T$] has unique solution for x, provided $2h_1 + h_2 = 0$.

28 Find the inverse of the following matrices

(a) $A = \begin{bmatrix} -2 & 2 \\ 3 & 4 \end{bmatrix}$ (b) $A = \begin{bmatrix} 2 & 1 & 1 \\ 1 & 2 & 1 \\ 1 & 1 & 2 \end{bmatrix}.$

29 Prove rank $AB \leq$ rank A and rank $AB \leq$ rank B, for any $A \in F^{m,n}$, $B \in F^{n,p}$. (Use Exercise 24, and the fact that row rank = col rank, p. 223.)

30 Let V, W be vector spaces over F, of dimensions n, m, respectively. Prove that $\mathrm{Hom}_F(V, W) \cong F^{m,n}$ (isomorphism of vector spaces; for definition of $\mathrm{Hom}_F(V, W)$ see Exercise 17). Hence show that dim $\mathrm{Hom}_F(V, W) = mn$.

31 Find the matrix A of the map $f: F^m \to F^n$ of Exercise 18, with respect to the bases of unit vectors of F^m, F^n.

32 *Orthogonal matrices.* An $n \times n$ matrix over R is said to be *orthogonal* if $A^T A = I_n$. Prove that the set $0_n(R)$ of all orthogonal $n \times n$ matrices over R is a subgroup of $GL_n(R)$. Prove that the group $S = 0_2^+(R) \cup 0_2^+(R) . J$ defined in Example 287 (p. 234) is the same as the group $0_2(R)$ as just defined.

33 If $x = (x_1, \ldots, x_n)^T \in R^{n,1}$, one defines the *length of* x to be the real number $|x| = \sqrt{x_1^2 + \cdots + x_n^2}$. Prove that if $A \in 0_n(R)$, then the map $f_A: R^{n,1} \to R^{n,1}$ 'preserves lengths', i.e. $|x| = |f_A(x)|$, for all $x \in R^n$.

Tables†

1. Symmetric group S(3)

ι	α	β	ρ	σ	τ
α	β	ι	τ	ρ	σ
β	ι	α	σ	τ	ρ
ρ	σ	τ	ι	α	β
σ	τ	ρ	β	ι	α
τ	ρ	σ	α	β	ι

$S(3)$ is the group of all permutations of the set $\{1, 2, 3\}$. Written in cycle notation (p. 35), the elements of $S(3)$ are $\iota = \mathbf{1}$ (identity map), $\alpha = (123)$, $\beta = (132)$, $\rho = (23)$, $\sigma = (13)$, $\tau = (12)$.

---

†In these multiplication tables, the unit element is always put in the top left-hand corner; the product xy is the element in row x, column y.

2. Cyclic group of order 6

1	x	x^2	x^3	x^4	x^5
x	x^2	x^3	x^4	x^5	1
x^2	x^3	x^4	x^5	1	x
x^3	x^4	x^5	1	x	x^2
x^4	x^5	1	x	x^2	x^3
x^5	1	x	x^2	x^3	x^4

This is the subgroup of the symmetric group S(6) which is generated by the cycle $x = (123456)$. Notice that $x^6 = 1$, hence $x^{-1} = x^{-1}x^6 = x^5$, $x^{-2} = x^4$, $x^{-3} = x^3$, etc.

Alternative interpretations of this group

(1) As the subgroup of the multiplicative group C^* of the complex field C (see p. 122) which is generated by the complex number

$$x = \cos\frac{2\pi}{6} + i\sin\frac{2\pi}{6} = \frac{1}{2} + \frac{\sqrt{3}}{2}i.$$

(2) As the subgroup of $GL_2(R)$ (see section 9.14) which is generated by the matrix

$$x = \begin{bmatrix} \cos\dfrac{2\pi}{6} & -\sin\dfrac{2\pi}{6} \\ \sin\dfrac{2\pi}{6} & \cos\dfrac{2\pi}{6} \end{bmatrix}$$

(this matrix a rotation of the plane through angle $2\pi/6$, see Example 285).

3. Alternating group A(4)

1	*t*	*u*	*v*	*a*	*b*	*c*	*d*	*p*	*q*	*r*	*s*
t	**1**	*v*	*u*	*b*	*a*	*d*	*c*	*q*	*p*	*s*	*r*
u	*v*	**1**	*t*	*c*	*d*	*a*	*b*	*r*	*s*	*p*	*q*
v	*u*	*t*	**1**	*d*	*c*	*b*	*a*	*s*	*r*	*q*	*p*
a	*c*	*d*	*b*	*p*	*r*	*s*	*q*	**1**	*u*	*v*	*t*
b	*d*	*c*	*a*	*q*	*s*	*r*	*p*	*t*	*v*	*u*	**1**
c	*a*	*b*	*d*	*r*	*p*	*q*	*s*	*u*	**1**	*t*	*v*
d	*b*	*a*	*c*	*s*	*q*	*p*	*r*	*v*	*t*	**1**	*u*
p	*s*	*q*	*r*	**1**	*v*	*t*	*u*	*a*	*d*	*b*	*c*
q	*r*	*p*	*s*	*t*	*u*	**1**	*v*	*b*	*c*	*a*	*d*
r	*q*	*s*	*p*	*u*	*t*	*v*	**1**	*c*	*b*	*d*	*a*
s	*p*	*r*	*q*	*v*	**1**	*u*	*t*	*d*	*a*	*c*	*b*

A(4) is the set of all even permutations of $\{1, 2, 3, 4\}$ (see Example 152, p. 93). Written in cycle notation the elements of A(4) are **1** (identity map), $t = (12)(34)$, $u = (13)(24)$, $v = (14)(23)$, $a = (123)$, $b = (243)$, $c = (142)$, $d = (134)$, $p = (132)$, $q = (143)$, $r = (234)$, $s = (124)$.

4. The 'fours group' V

1	t	u	v
t	1	v	u
u	v	1	t
v	u	t	1

This is a subgroup of A(4). Notice that V is in fact a normal subgroup of A(4) (and of S(4), which contains A(4)). The three cosets of V in A(4) are: $V = \{1, t, u, v\}$, $A = Va = \{a, b, c, d\}$, $B = Vp = \{p, q, r, s\}$. See Example 137, p. 84.

5. The dihedral group D(8)

1	a	a^2	a^3	b	ab	a^2b	a^3b
a	a^2	a^3	1	ab	a^2b	a^3b	b
a^2	a^3	1	a	a^2b	a^3b	b	ab
a^3	1	a	a^2	a^3b	b	ab	a^2b
b	a^3b	a^2b	ab	1	a^3	a^2	a
ab	b	a^3b	a^2b	a	1	a^3	a^2
a^2b	ab	b	a^3b	a^2	a	1	a^3
a^3b	a^2b	ab	b	a^3	a^2	a	1

This can be regarded as the subgroup of S(4) which is generated by the permutations $a = (1234)$, $b = (24)$. To check the table, first show that

$$a^4 = 1, \qquad b^2 = 1, \qquad ba = a^3b. \tag{1}$$

It will be found that all the products in the table follow from these 'relations' (1). For example: $ab \cdot a^2b = abaab = aa^3bab = aa^3a^3bb = a^7b^2 = a^3$, using $ba = a^3b$ to move b to the right. In cycle notation the elements are $1 =$ identity map, $a = (1234)$, $a^2 = (13)(24)$, $a^3 = (1432)$, $b = (24)$, $ab = (12)(34)$, $a^2b = (13)$, $a^3b = (14)(23)$. Notice that this group contains V as a subgroup.

Alternative interpretation

We can regard this group as the subgroup of $GL_2(R)$ generated by the matrices

$$a = \begin{bmatrix} 0 & -1 \\ 1 & 0 \end{bmatrix}, \qquad b = \begin{bmatrix} 1 & 0 \\ 0 & -1 \end{bmatrix}.$$

Check that these satisfy the relations (1). (This is case $n = 4$ of Example 288, p. 234; we have $a = A(2\pi/4)$, $b = J$.)

6. Quaternion group Q(8)

I	$-I$	P	$-P$	Q	$-Q$	R	$-R$
$-I$	I	$-P$	P	$-Q$	Q	$-R$	R
P	$-P$	$-I$	I	R	$-R$	$-Q$	Q
$-P$	P	I	$-I$	$-R$	R	Q	$-Q$
Q	$-Q$	$-R$	R	$-I$	I	P	$-P$
$-Q$	Q	R	$-R$	I	$-I$	$-P$	P
R	$-R$	Q	$-Q$	$-P$	P	$-I$	I
$-R$	R	$-Q$	Q	P	$-P$	I	$-I$

This is the subgroup of $GL_2(C)$ whose elements are $\pm I$, $\pm P$, $\pm Q$, $\pm R$;

$$I = \begin{bmatrix} 1 & 0 \\ 0 & 1 \end{bmatrix}, \qquad P = \begin{bmatrix} 0 & 1 \\ -1 & 0 \end{bmatrix},$$

$$Q = \begin{bmatrix} i & 0 \\ 0 & -i \end{bmatrix}, \qquad R = \begin{bmatrix} 0 & -i \\ -i & 0 \end{bmatrix}.$$

See Examples 268 (p. 203) and 289 (p. 235).

To check the table first verify the relations

$$P^2 = Q^2 = R^2 = -I$$

and

$$PQ = R, \qquad RQ = P, \qquad QP = R.$$

All the products in the table above follow from these relations.

List of Notations

G/H	quotient group, p. 83; quotient ring, p. 136; quotient space, p. 220
$U(S)$	set (or group) of invertible elements of ring S, p. 113
$S[X]$	set (or ring) of polynomials in X over the commutative ring S, p. 125
$\deg f$	degree of polynomial f, p. 127
Sb	principal ideal generated by element b in commutative ring S, p. 139
$b \vert a$	b divides a, p. 148
$a \sim b$	a is associated to b, p. 148
hcf	highest common factor, p. 152
PID	principal ideal domain, p. 140
F^n	set of all n-tuples (p. 11) i.e. *n-vectors* $x = (x_1, \ldots, x_n)$ over field F (p. 169)
$L(a, b)$	line through points $a, b \in R^n$, p. 174
sp S	subspace spanned by subset S of vector space, p. 176
$L(a, b, c)$	plane through points $a, b, c \in R^n$, p. 178
dim V	dimension of vector space V, p. 185
$F^{m,n}$	set of all $m \times n$ matrices over F, p. 194
A^T	transpose of matrix A, p. 199
row sp A	row space of matrix A, p. 204
col sp A	column space of matrix A, p. 204
$M_n(F)$	ring of $n \times n$ matrices over F, p. 201
$GL_n(F)$	group of invertible $n \times n$ matrices over F, p. 227
R	set of all real numbers, p. 2
C	set of all complex numbers, p. 122
Z	set of all integers, p. 2

Answers to exercises

Chapter 1

5 Use Exercise 4 to show $x = |A_1 \cup A_2 \cup A_3| = 28 - a_{13} + a_{123}$. But $a_{13} \geq a_{123}$ (since $A_1 \cap A_3 \supseteq A_1 \cap A_2 \cap A_3$) hence $x \leq 28$. Also $|A_2 \cup A_3| = a_2 + a_3 - a_{23} = 26$, hence $x \geq 26$.

9 No. If $x \in A - B$, then $(x, x) \in (A \cup B) \times (A \cup B)$, but $(x, x) \notin (A \times B) \cup (B \times A)$.

Chapter 2

1 (i) Eq. (ii) Sym. (iii) Ref., sym. (iv) Ref., tra.

5 (i) $E_0 = \{\ldots, -2, 0, 2, 4, \ldots\}$, $E_1 = \{\ldots, -1, 1, 3, 5, \ldots\}$
   (ii) $E_0 = \{\ldots, -3, 0, 3, 6, \ldots\}$, $E_1 = \{\ldots, -2, 1, 4, 7, \ldots\}$, $E_2 = \{\ldots, -1, 2, 5, 8, \ldots\}$.

6 $2^4 \equiv 1 \bmod 5$, so $2^{512} = (2^4)^{128} \equiv 1^{128} \equiv 1 \bmod 5$.

8 $A \cap B$ is congruence class of 7 mod 12.

9 (i) Since $10 \equiv -1 \bmod 11$, $x \equiv b_0 - b_1 + b_2 - \cdots \bmod 11$.
   (ii) Since $10 \equiv -2 \bmod 12$, $x \equiv b_0 - 2b_1 + 4b_2 - 8b_3 + 16b_4 - \cdots \equiv b_0 - 2b_1 + 4b_2 + 4b_3 + 4b_4 + \cdots \bmod 12$.

Chapter 3

1 6 are injective. $s(s-1)(s-2) \ldots (s-r+1)$.

2 (i) neither, (ii) both, (iii) surj., (iv) inj.

3 $\theta^{-1}(y) = \sqrt{y}$. $\phi(x) = \phi(-x)$, so ϕ not bij.

4 $\theta^{-1}(y) = a^{-1}(y - b)$.

5 Commute if and only if $ad + b = bc + d$.

6 No, e.g.

$$\delta = \begin{pmatrix} 1 & 2 & 3 \\ a & b & a \end{pmatrix}, \qquad \phi = \begin{pmatrix} a & b \\ 1 & 2 \end{pmatrix}.$$

7 $\alpha = (123)$, $\rho = (23)$, $\alpha\rho = (12)$, $\rho\alpha = (13)$.

8 (a) (bfdg) (ce).

11 $\theta: R \to R$ (Ex. 48) is inj., not surj. Map $\phi: Z \to Z$ given by $\theta(x) = [\frac{1}{2}x]$ (Exercise 3, Chapter 2) is surj. not inj. Not possible if A finite.

12 Use map $\theta(x) = \frac{1}{2}x$ (x even).

13 Use maps (i) $(a, b) \to (b, a)$, (ii) $[(a, b), c] \to [a, (b, c)]$.

14 Take bijection $f: A \to B$. Now define $g: S(A) \to S(B)$ by $g(\theta) = f\theta f^{-1}$, and prove g is bijective.

Chapter 4

1 There are n^{n^2} ways of filling all n^2 places in an $n \times n$ multiplication table with elements of A. (i) 8; (ii) 8; (iii) 4; (iv) 4.

2 Neither.

7 Not unless $|A| = 1$.

9 $(\alpha\rho)^2 = 1$, $\alpha^2\rho^2 = (132)$.

11 Show $\hat{x} = -x/(1+x)$ is inverse of $x \in R'$ with respect to $\circ$.

13 $Z^* = \{1, -1\}$, $|Z^*| = 2$.

15 $(\theta_{a,b})^{-1} = \theta_{c,d}$, where $c = a^{-1}$ and $d = -a^{-1}b$. Unit element is $\theta_{1,0}$.

Chapter 5

1 Yes.

2 One way is to arrange that $1 \circ 1 = 2$, $2 \circ 2 = 3, \ldots, n \circ n = 1$ (other products arbitrary).

4 $H^* = \{x/y \,|\, x, y \text{ non-zero integers}\}$.

9 Prove any two words in X (p. 60) commute.

10 Since m, n are both multiples of d, they are in $\mathrm{gp}\{d\}$; hence $\mathrm{gp}\{m, n\} \subseteq \mathrm{gp}\{d\}$. For converse inclusion use theorem (i) section 8.13 (p. 152).

13 e has order 1; t, u, v have order 2; others have order 3.

15 If $x \in G$, $\mathrm{gp}\{x\}$ is finite, hence $n = |\mathrm{gp}\{x\}| = \text{order } x$ is finite. So if $x \in H$, $x^{-1} = x^{n-1} \in H$ by S1.

16 $C(\iota) = S(3)$; $C(\alpha) = C(\beta) = \{\iota, \alpha, \beta\}$; if $\xi = \rho$, σ or τ then $C(\xi) = \{\iota, \xi\}$.

17 Normalizer of $\{\beta, \beta^{-1}\}$ is group $D(8)$ of Table 5 (p. 244).

19 $xH = H \Leftrightarrow x \in H$ (see Example 116, section 6.1, p. 69).

Chapter 6

1 Right H, Ht, Hu, Hv; left H, tH, uH, vH.

2 The one-element subsets $\{g\}$, $g \in G$. Only G itself.

3 $y \in xH \cap xK \Leftrightarrow x^{-1}y \in H \cap K \Leftrightarrow y \in x(H \cap K)$.

4 Each Hg contains one element of X, so each $g^{-1}H = (Hg)^{-1}$ contains one element of X^{-1}.

6 If $\{x_\mu | \mu = 1, \ldots, m\}$ is rt. trans. H in K, and $\{y_\nu | \nu = 1, \ldots, n\}$ is a rt. trans. K in G, prove $\{x_\mu y_\nu\}$ is a rt. trans. H in G.

8 $z \in \text{Stab}(Hx) \Leftrightarrow xzx^{-1} \in H$ (by theorem section 6.1) $\Leftrightarrow z \in xHx^{-1}$.

9 $O_f = \{x_1x_2 + x_3x_4, x_1x_3 + x_2x_4, x_1x_4 + x_2x_3\}$. $H = V \cup Vy$, where $V = \{1, t, u, v\}$ (see Table 3, p. 244) and $y = (12)$.
$|H| = 8$, so H is a Sylow two-subgroup of $S(4)$.

10 C. classes $D(8)$: $\{1\}, \{a^2\}, \{a, a^3\}, \{b, a^2b\}, \{ab, a^3b\}$; centre $\{1, a^2\}$. C classes $Q(8)$: $\{I\}, \{-I\}, \{P, -P\}, \{Q, -Q\}, \{R, -R\}$; centre $\{I, -I\}$.

11 $[1^4], [1^22], [13], [4], [2^2]$.

12 $\{1, \alpha, \alpha^2, \ldots, \alpha^{n-1}\}$.

13 $(n-1)!$

14 $\{e\}, \{t, u, v\}, \{a, b, c, d\}, \{p, q, r, s\}$.

Chapter 7

8 Let $G = \text{gp}\{x\}$. If G has order n, then subgroups of G are $\text{gp}\{x^d\}$ for every d dividing n. If d, d' both divide n, then $\text{gp}\{x^d\} \subseteq \text{gp}\{x^{d'}\}$ if and only if d' divides d.

9 Subgroups: $\{1\}$, $\{1, a^2\}$, $\{1, b\}$, $\{1, ab\}$, $\{1, a^2b\}$, $\{1, a^3b\}$, $\{1, a, a^2, a^3\}$, $\{1, a^2, b, a^2b\}$, $\{1, a^2, ab, a^3b\}$, $D(8)$.

10 Define $\psi: M \to G$ by $\psi(m) = \theta(m)$, $m \in M$. Apply Homo. Thm. to ψ.

11 $M \trianglelefteq F \Leftrightarrow fM = Mf$, all $f \in F \Leftrightarrow \theta(f)H = H\theta(f)$, all $f \in F \Leftrightarrow H \trianglelefteq G$ (since θ is epi.). Prove map $fM \to \theta(f)H$ is iso. $F/M \to G/H$.

12 If S is subgroup of G, then $S = S^{-1}$. So $UV = (UV)^{-1} = V^{-1}U^{-1} = VU$.

17 If x, y, z are generators of C_m, C_n, C_{mn}, define $f: C_m \times C_n \to C_{mn}$ by $f(x^r, y^s) = z^{rn+sm}$. Prove this is iso.

18 Let $(a, b) \in L$, $(a, b) \neq (0, 0)$, then $L = \{(\lambda a, \lambda b) | \lambda \in R\}$ is line through $(0, 0)$ and (a, b). Check L is subgp. Coset $(x, y) + L$ is line through (x, y), parallel to L.

Chapter 8

1 $U(Z/5Z) = \{E_1, E_2, E_3, E_4\} = \text{gp}\{E_2\}$.

5 E_0, E_1, E_3, E_4.

8 Remember F is commutative.

9 If $v = z_0 + z_1\sqrt{2}$ then $v\bar{v} = z_1^2$, where $\bar{v} = z_0 - z_1\sqrt{2}$. If $v \neq 0$ then $r = z_0^2 - 2z_1^2 \neq 0$ (since $z_0^2 - 2z_1^2 = 0$ with $z_0, z_1 \neq 0$ leads to $\sqrt{2} = \pm z_0/z_1$, impossible since $\sqrt{2}$ is not rational) hence $v^{-1} = r^{-1}z_0 - r^{-1}z_1\sqrt{2}$. So every uv^{-1} ($u, v \in \text{rg}\{\sqrt{2}\}$ has form $r_0 + r_1\sqrt{2}$, r_0, r_1 rational.

12 $1, -1 - 3i, \frac{1}{10}(-1 + 3i), \frac{1}{10}(-1 + 3i); u^{-1} = \frac{1}{5}(2 - i), v^{-1} = \frac{1}{2}(-1 + i)$.

14 $f = 1 - X + 2X^2$.

15 Assume $f \neq 0$. Since $\deg f^2 = 2 \deg f = \deg f$, we must have $\deg f = 0$, i.e. $f = f_0 \in S$. Then $f_0^2 = f_0 \Rightarrow f_0^2 - f_0 = 0 \Rightarrow f_0(f_0 - 1) = 0$, hence $f_0 = 0$ or $f_0 - 1 = 0$, since S is int. dom.

17 By def. $\theta(1) = 1$. Hence $\theta(m) = m$, all integers m; hence $\theta(m/n) = \theta(m \cdot n^{-1}) = \theta(m)\theta(n)^{-1} = m/n$, all $m, n \in Z$ ($n \neq 0$).

18 Yes.

19 Let $z = \theta(\frac{1}{2})$, then $2z = \theta(1) = 1$, impossible for $z \in Z$.

20 Take $0 \neq x \in S$. Then xS is non-zero ideal of S, hence $xS = S$ because S simple. So there is $y \in S$. $xy = 1$, i.e. x has inverse in S.

23 Max. ideals of Z are pZ, p prime.

24 Use Exercise 22.

30 Quotient $X^3 - 2X$, remainder $2X + 1$.

31 $h(X) = X^2 - X - 1$.

33 Order n of an element x in finite group G, divides order g of G since r is the order of a subgroup $H = \text{gp}\{x\}$ of G (use Lagrange). So $x^g = 1$, all $x \in G$. Apply this to $G = U(F)$.

35 Use Example 213, p. 147. Note $X^2 + aX + b = X^2 - 2a_0X + (a_0^2 + a_1^2) \Rightarrow a^2 = 4a_0^2 < 4(a_0^2 + a_1^2) = 4b$, if $a_1 \neq 0$.

36 $X, X + 1, X^2 + X + 1, X^3 + X + 1, X^3 + X^2 + 1, X^4 + X^3 + 1, X^4 + X + 1, X^4 + X^3 + X^2 + X + 1$.

37 hcf $21 = 2.420 - 3.273$.

38 [19].

39 hcf $X^2 + X + 1 = -X(X^2 + X + 1)a(X) + b(X)$.

40 Write $a = X + (X^2 + X + 1) F[X]$; elements are $z_0 + z_1 a$ (z_0, z_1 integers mod 2); note $a^2 + a = 1$.

0	1	a	$1 + a$
1	0	$1 + a$	a
a	$1 + a$	0	1
$1 + a$	a	1	0

1	a	$1 + a$
a	$1 + a$	1
$1 + a$	1	a

Addition   *Multiplication*

Chapter 9

4 $L(a, b) = \{(-1 + 3\lambda, 4 - 3\lambda)|\lambda \in R\}$, $L(c, d) = \{(1 - 3\mu, 3 - \mu)|\mu \in R\}$.
   Intersection $\frac{1}{4}(1, 11)$.

5 $l = \frac{1}{2}(a + b)$, $m = \frac{1}{2}b$, $n = \frac{1}{2}a$. Point $x = \frac{1}{3}(a + b) = \frac{1}{3} \cdot 0 + \frac{2}{3}l = \frac{1}{3}a + \frac{2}{3}m = \frac{1}{3}b + \frac{2}{3}n$ lies on $L(0, l)$, $L(a, m)$ and $L(b, n)$.

6 Show $(1 - \lambda)u + \lambda v$ lies on Π_α, all $\lambda \in R$.

7 $\Lambda = L[(0, 0), (-b, a)]$.

8 Take any $u = (u_1, u_2)$ such that $au_1 + bu_2 = h$.

9 T_0 is identity el. of $S(R^n)$; $T_x T_y = T_{x+y}$, $(T_x)^{-1} = T_{-x}$.

13 dim $\Pi = 2$.

14 (For example) $x = (1, 1, 0)$, $y = (0, 1, 1)$; dim $\Pi = 2$.

18 Ker $f = \{0, \ldots, x_{m+1}, \ldots, x_n)|x_{m+1}, \ldots, x_n \in F\}$, Im $f = F^n$.

20 $\mathscr{S}$ has basis $\{E_{ij} + E_{ji}|1 \leq i \leq j \leq n\}$, $\mathscr{A}$ has basis $\{E_{ij} - E_{ji}|1 \leq i < j \leq n\}$.

24 If $b_{(j)} = b_{j1}, b_{j2}, \ldots, b_{jp})$, and $c_{(i)} = (c_{i1}, c_{i2}, \ldots, c_{ip})$ then $c_{(i)} = a_{i1}b_{(1)} + a_{i2}b_{(2)} + \cdots + a_{in}b_{(n)}$.

25 3.

26 $2h_1 - h_2 - h_3 + h_4 = 0$. A basis is $\{(-4, -1, 4, 1)\}$.

27 2.

28 (a) $-\frac{1}{14} \begin{bmatrix} 4 & -2 \\ -3 & -2 \end{bmatrix}$ (b) $\frac{1}{4} \begin{bmatrix} 3 & -1 & -1 \\ -1 & 3 & -1 \\ -1 & -1 & 3 \end{bmatrix}$

30 Use map $\theta : \text{Hom}(V, W) \to F^{m,n}$ which takes $f \in \text{Hom}(V, W)$ to matrix $M_{\mathbf{v},\mathbf{w}}(f)$, where $\mathbf{v}$, $\mathbf{w}$ are bases of V, W.

31 A is $n \times m$ matrix $\begin{bmatrix} 1 & 0 & 0 & 0 & \cdots & 0 \\ 0 & 1 & 0 & 0 & \cdots & 0 \\ \vdots & \vdots & \vdots & \vdots & & \vdots \\ 0 & 0 & 1 & 0 & \cdots & 0 \end{bmatrix}$ $\vdots$

33 $|x|^2 = x_1^2 + \cdots + x_n^2 = x^T x$,   and $|f_A(x)|^2 = |Ax|^2 = (Ax)^T Ax = x^T A^T Ax = x^T I_n x = x^T x = |x|^2$.

Index